SCIENCE 101

BIOLOGY

Produced for HarperCollins by:

Hydra Publishing
129 Main Street
Irvington, NY 10533
www.hylaspublishing.com

FIRST EDITION

The name of the "Smithsonian," "Smithsonian Institution," and the
sunburst logo are registered trademarks of the Smithsonian Institution.

Library of Congress Cataloging-in-Publication Data

Ochoa, George.
  Science 101: Biology / George Ochoa. -- 1st ed.
      p. cm.
  Includes bibliographical references and index.
  ISBN: 978-0-06-089135-0
  ISBN-10: 0-06-089135-1
  1. Biology. I. Title. II. Title: Science one hundred one.

  QH308.2.O24 2007
  570--dc22

                              2007045474

07   08   09   10   QW   10 9 8 7 6 5 4 3 2 1

SCIENCE 101

BIOLOGY

George Ochoa

Collins

An Imprint of HarperCollins*Publishers*

CONTENTS

WELCOME TO BIOLOGY

Left: Slime mold spores magnified by electron microscope. Top: Bee on flower. Above: Venus flytrap. Life is amazing for its diversity. Its manifestations include slime molds, protists that resemble fungi and reproduce by spores carried in the wind; bees, insects that pollinate flowers as they gather food from them; and Venus flytraps, plants that eat insects for their nitrogen. Biology is the study of all organisms, no matter their size or shape.

Living things tend to fascinate people of all ages. Infants are mesmerized by the family dog, toddlers by the scurrying of insects. Schoolchildren are amazed by the monarch butterfly emerging from its cocoon; adolescents gaze in the mirror, seeing their bodies develop. Adults cultivate gardens, watch birds, hike through forests. And still the wonders of life never cease. A person could spend a lifetime investigating and observing them. Some people do. They are biologists, and their life's work is the study of living things.

Biology is a study that is never completed; there always is more to learn. Nobody is certain how many species of living things there are. Estimates suggest at least 10 million, but there may be many more. Living things are incredibly diverse: There are organisms that consist of just one cell, such as bacteria. Other kinds have many cells but are tiny, such as slime molds. Some multicelled creatures, like the blue whale, are enormous. Organisms are complex, with structures as powerful as an eagle's wings and as subtle as the alveoli, the minuscule air chambers in the lungs. The interactions among organisms, such as bees pollinating flowers even as they gather food, are equally amazing.

BIOLOGISTS AT WORK

In their quest to understand living things, scientists employ a variety of tools and techniques. To assist them in observing nature, they use specialized equipment like microscopes, petri dishes, dissection kits, gene sequencers, satellite collars, and submersibles. To help them test their hypotheses, or tentative conclusions, they set up ingenious experiments. In the seventeenth century, Italian scientist Francesco Redi placed meat in containers—some covered, others uncovered—to demonstrate that maggots were not generated spontaneously from decaying meat but hatched from the eggs of flies.

Since Redi's time, biologists have learned a great deal about how living things work. Darwin explained how life evolved; Mendel discovered the laws of genetics; Pasteur showed that germs cause disease; and Watson and Crick revealed the structure of DNA. Today biologists in numerous subdisciplines—among them microbiology, biochemistry, botany, zoology, anatomy, physiology, ecology, marine biology, and neuroscience—build on past discoveries and attempt to penetrate the mysteries that yet remain. Even though much still is unknown, a great deal of knowledge has accumulated. Indeed, so much is known that no individual specialist can grasp everything that has been discovered in every field.

THE COMPLEXITIES OF LIFE

The basic characteristics that distinguish life from nonlife are fairly clear. These characteristics include reproduction, growth, and metabolism. Another basic characteristic is the concept that living things exhibit increasingly complex levels of organization. Subatomic particles make up atoms, which make up molecules, which make up cells—the basic unit of all living things. But as complexity increases, it becomes harder to trace the details of every organism's functioning. In a typical animal, cells are the building blocks of tissues, organs, and the systems in which the organs operate. Within the animal, the individual physiological structures (lungs, blood vessels, intestines) carry out specific functions (respiration, circulation, digestion). These functions enable the survival and reproduction of the organism.

Organisms themselves are parts of still higher levels of organization, such as populations and ecosystems. Today's organisms evolved from simpler forms and are related to one

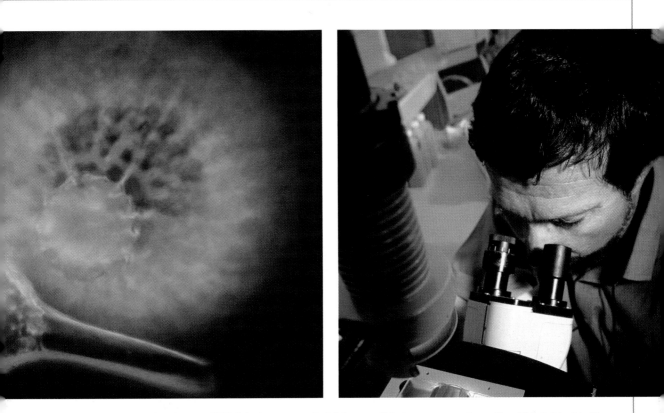

Above left: Radiolaria under microscope. With their glassy, symmetrical skeletons, radiolarians are among the most beautiful of all protists. They are also an ancient group, dating to the early Cambrian Period. Above right: Biological researcher uses a light microscope to study oyster toxins. Microscopes are among the most important tools biologists use. With microscopes, they can see organisms that are too small for the naked eye, such as the one-celled radiolaria. They also can see microscopic features of larger organisms, such as the evidence of toxins in oysters.

another through the elaborate branchings of the evolutionary tree of life. Humans, too, are part of that tree, though the mysterious workings of their larger brains make them more complex than their primate cousins.

BIOLOGY MADE CLEAR

Given the complexity of living things and the quantity of knowledge that exists about them, there is a compelling need for a clear understanding of at least the basics of biology. *Science 101: Biology* aims to provide that understanding. It is designed to guide the general reader through the wonderfully complex mysteries of the science of life. In 12 chapters, this book explores how biologists

> Biology is a study that is never completed; there is always more to learn.

define life. It sheds light on their work and accomplishments. It helps the general reader navigate the world of cells; anatomy; physiology; ecology; evolution; classification; plants and animals; humans; and the latest biological findings and controversies.

Today more than ever, biology plays a vital role in society. Genomic discoveries become the basis of new medical treatments; ecological research shows us how to improve agriculture and preserve species. Above all, biology is important because it helps satisfy the human sense of wonder about the living world. *Science 101: Biology* builds a solid foundation of understanding to support that sense of wonder.

CHAPTER 1

THE MYSTERY OF LIFE

Top: Canada goose in flight. Bottom: Moss covering rock. Left: California redwood forest, home to the world's largest trees. The mystery of life, which includes a wide-range of phenomena, from the flight of birds to the magnificent heights of the California redwoods, spurs biological inquiry. A typical question a biologist might ask may be as simple as: What makes moss different from the rock that it covers?

Almost every aspect of life is mysterious. How do birds fly? What makes a heart beat? Why are redwood trees tall, and why does a child resemble her parents? How does disease spread? Why are humans different from other animals? What makes the moss on a rock different from the rock? How did living things get here? What is life?

A good way to begin answering questions like these is to understand some of the basic concepts of biology. Biologists have learned to recognize key characteristics that often are found in living things, including reproduction, growth, metabolism, and adaptation, to answer the question, "What is life?" Living things also can be understood as concentric layers of organization, from subatomic particles to the biosphere itself, the total region of Earth where life exists. It is also important to understand persistent themes in the history of life. These include continuity and change (evident in the process known as evolution) and a strong relationship between structures and the functions they perform. Such concepts are a useful beginning to probing the mystery of life.

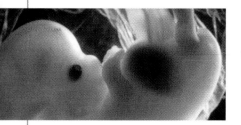

What Is Life?

Life is not easy to define, even for biologists, the people who study it as a profession. The average person can usually tell a living thing from a nonliving one, but articulating what makes them different is difficult. Is the difference that an ant moves but a pebble does not? No: a pebble moves under certain conditions, such as rolling down a slope. Is it that a flower grows but a crystal doesn't? But crystals do grow under the right conditions. There may not be a comprehensive one-line definition of life, but life can be defined in terms of several characteristics that are typically present in living things.

These include reproduction, growth, metabolism, irritability, and adaptation.

REPRODUCTION AND GROWTH

Living things make more of themselves—in two senses. Through reproduction, they make more individuals of their own species, or kind. Through growth, they make their own bodies larger in an orderly way, with all parts increasing in size. Both processes depend on division of the most basic biological unit: the cell. A cell splits to form two new cells. In single-celled organisms, such as bacteria, this is enough to

bring about reproduction: two organisms now exist instead of one. In organisms that have many cells and reproduce sexually, the process is more complicated. Simple cell division will result in growth of tissues—for example, when the muscle cells in a weightlifter's arms divide, the arms thicken. But reproduction requires the union of two sex cells, one each from a male and a female. The result is an embryo that will grow into a new individual.

USING AND FINDING FOOD

To stay alive, living things must take in energy and raw materials, use them, and discard the waste

Bottom left: Human cell dividing. Bottom right: Bee collecting pollen. Top left: Human embryo, seven weeks.

An insect's compound eyes.

An elephant ear sponge, like all sponge species, possesses the basic characteristics of life.

products. They get their energy in many ways—a meadow absorbing sunlight, bees collecting pollen and nectar from flowers, or a tiger eating its prey. But they all need the energy and raw materials known as food to fuel their reactions, maintain their bodies, and grow. The sum of the chemical processes by which organisms perform these activities is known as metabolism.

To succeed in finding food—and in the equally important task of avoiding becoming food—living things must be responsive to their environments. They must have the property of irritability, or sensitivity, the ability to sense and respond to stimuli. The sense apparatus may be as simple as the light-sensitive eyespot of a one-celled algae or as complex as an insect's compound eye. Sensitivity usually sparks some kind of movement, such as the sudden flight of birds when a person draws near, or the bending of a plant toward sunlight.

WHY A SPONGE IS ALIVE

The household sponge, used for everything from washing dishes to cleaning cars, is clearly not alive. But what about the sponge found in the ocean? A porous mass attached to rocks, it does not move from place to place and has no internal organs. Why should it qualify as a living thing? Because it shows the basic characteristics of life.

First, it reproduces, sexually or asexually. A sponge not only grows but can also regenerate lost parts of its body. It has a metabolism: its cells are able to absorb and use food. It cannot move from its perch, but certain of its cells have threadlike structures called flagella that whip about, creating water currents that bring food into its pores. It can stop the water flow in response to stimuli, demonstrating irritability. Among the world's oldest animals, sponges are well adapted to their undersea environment—and definitely alive.

To survive, living things must also fit themselves, or adapt, to their environments. They do so in the long term when the most successful members of a population—the fastest sharks or the tallest oaks—pass on to their descendants those characteristics that made them successful. This kind of adaptation contributes to the process of evolution through which species change and new species originate. Adaptation happens in the short term when an organism adjusts to passing changes, as when a person's pupils shrink on leaving a dark room and entering into sunlight. This and other adaptations are part of the arsenal that living things draw upon in order to metabolize, grow, and reproduce—the basic characteristics of life.

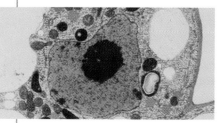

A Tendency Toward Order

Although life is different from nonlife, it is not completely different. Living things exist in a nonliving universe and depend on it in many ways, from plants absorbing energy from sunlight to bats finding shelter in caves. Indeed, living things are made of the same tiny particles—subatomic particles—that make up nonliving things. What makes organisms different from the materials that compose them is their level of organization.

Living things exhibit not just one but many layers of biological organization. This tendency toward order is sometimes modeled in a pyramid of life.

THE PYRAMID OF LIFE

In this pyramid, each level has structures that are larger and more complex than those below it. The structures on each level contain those below it, but do not contain any above it. For example, an organ contains tissues, which contain cells, but organs do not contain organisms; rather, it is the organism that contains organs. So the pyramid of life is a pyramid of increasing complexity until one reaches the top, the entire biosphere, or the region of Earth that contains life.

The pyramid levels are:

1. **subatomic particle**—A unit of matter, such as a proton, electron, or neutron that can compose atoms.

2. **atom**—A larger unit of matter that can compose even larger units called molecules.

3. **molecule**—A molecule is the smallest part of a substance that still has the chemical identity of the substance. For example, a water molecule still behaves like water, but if broken into its constituents, one oxygen and two hydrogen atoms, it will not.

4. **organelle**—A structure within a cell made up of molecules and having a specific task. For example, a lysosome breaks down food or foreign particles.

5. **cell**—The smallest structure in the pyramid of life that can be called alive. It contains organelles and uses them to pursue such biological activities as reproduction, growth, and metabolism.

6. **tissue**—In multicellular organisms, a collection of cells that are similar in structure and work together to perform a particular function. Xylem, for example, is a plant tissue that transports water and minerals from the soil upward from the roots.

7. **organ**—A part of an organism that joins two or more kinds of tissues to perform specific functions. The human heart, the organ that pumps blood, consists of muscle as well as other tissues.

8. **organ system**—A group of organs that carries out a major activity. For example, the human circulatory system, containing the heart, blood vessels, and other components, transports materials throughout the body.

9. **organism**—An individual living system, such as a bacterium or a whale, capable of basic biological activities such as reproduction, growth, and metabolism.

10. **population**—A group of organisms of the same species living in one area at one time. All the cats in London are its population of cats.

11. **community**—A group of animal and plant populations, such as lions, antelopes, grasses, and trees, living and interacting together in the same area.

12. **ecosystem**—A community and its nonliving or physical environment, including soil, water, air, climate, and energy.

13. **biosphere**—The entire region on or near Earth's surface in which life can be found. Comprising seas, caves, skies, and land, it is the sum of all Earth's ecosystems.

PYRAMID OF LIFE

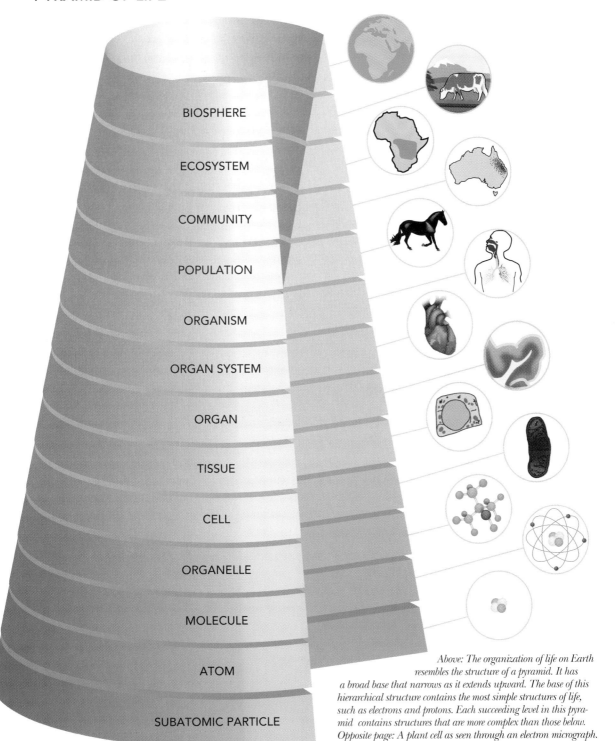

BIOSPHERE

ECOSYSTEM

COMMUNITY

POPULATION

ORGANISM

ORGAN SYSTEM

ORGAN

TISSUE

CELL

ORGANELLE

MOLECULE

ATOM

SUBATOMIC PARTICLE

Above: The organization of life on Earth resembles the structure of a pyramid. It has a broad base that narrows as it extends upward. The base of this hierarchical structure contains the most simple structures of life, such as electrons and protons. Each succeeding level in this pyramid contains structures that are more complex than those below. Opposite page: A plant cell as seen through an electron micrograph.

Constant Change

Earth is subject to constant change, but living things change differently than nonliving things do. For more than 400 million years, the Sahara area of Africa, which today is a sandy desert, was covered with glaciers. Yet even though the location of ice sheets has changed over that vast span of time, ice crystals today are exactly identical to ice crystals then. Not so for organisms: fish back then were jawless and armor-plated, the only land plants hugged the ground like liverworts, and there were no land animals at all.

Unlike mere physical entities, over time, living things can change radically in their basic structures and diversify into many new forms. Yet they also exhibit a striking continuity that allows biologists to trace the path by which the new forms developed. The reason that there are any land animals today is that certain sea animals adapted to living on land; some of these became the ancestors of all land vertebrates, including frogs, lizards, birds, bears, and humans. Despite the diversity of these organisms, all of them still bear a noticeable family resemblance to their fishy ancestors—from the backbone that makes them vertebrates to the antibodies that protect them against micro-

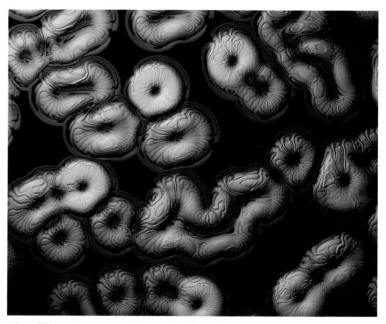

Above: This computer-generated image shows how chromosomes, coiled into compact struc-tures, during cell divsion make copies of themselves. Top left: Common liverwort. Over 400 million years ago, the only land plants were similar to modern liverworts. Plants diversified through evolutionary changes in their genes, units of heredity carried on chromosomes.

bial invaders. This mixture of continuity and change is made possible by two factors: genes and evolution.

GENES
Genes are units of heredity contained in an organism's cells and composed of DNA (deoxy-ribonucleic acid). When passed on to a descendant, the genes determine which characteristics of the parent the offspring will inherit, whether four legs, gills, yellow petals, or blue eyes.

In species that reproduce asexually, such as amoebas, the descendants share all the genes of the parent and are therefore their identical twins. In sexual reproduction the continuity is not so predictable: in each offspring, the genes of two parents are shuffled in a new combination.

EVOLUTION
The main engine of biological change is evolution. Through this process, entirely new categories

of life can emerge—from new species, such as Tyrannosaurus rex, to new kingdoms, such as animals. The principal mechanism of evolution is natural selection, which depends for its operation on random mutation.

Random mutations are sudden alterations in genetic material that naturally take place at a low rate and may happen faster in the presence of radiation and certain chemicals. Most mutations are harmful. Occasionally, however, one emerges that increases the fitness of an organism, that is, its potential to survive and reproduce: a mutation that allows a root to grow deeper or a nose to smell more distant prey are two examples. When the mutated gene is advantageous, it is likely to spread in the population because those who possess the gene out-reproduce those who do not. As such, naturally

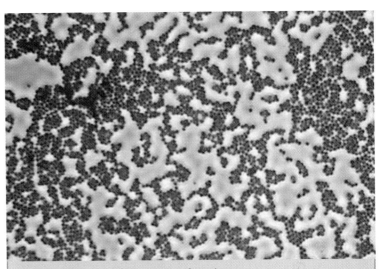

Staphylococcus aureus, *a resistant bacterium.*

ANTIBIOTICS AND BACTERIAL EVOLUTION

The discovery of antibiotics, which first came into widespread use in the United States during the 1940s, was a medical milestone. Many patients who once would have died of diseases such as pneumonia, tuberculosis, meningitis, and scarlet fever were saved when antibiotics destroyed the bacteria causing their illnesses. By the 1990s, however, the power of these wonder drugs was waning.

Certain strains of bacteria had emerged that were resistant to many antibiotics. After just five decades of natural selection, they had developed resistance. Some bacteria had been more resistant than others to antibiotics, and these survived and passed on that trait to their descendants, creating tough new strains that seemed to defy treatment.

Today, doctors are continuing to see increasing rates of resistance among bacteria such as *Staphylococcus aureus* and *Klebsiella pneumoniae.* Public health education, novel drugs, and controls on antibiotic use may help. But whatever weapons society develops, bacteria will continue to evolve in response.

Sarah van Fleet rose. As are many garden flowers, this rose is the product of human-directed selective breeding.

selected changes accumulate over time, and living things gradually evolve to be better suited for their environments. They also diversify to fill different niches, or environmental roles.

Living things exhibit continuity and change even on the level of the individual organism. The frog egg hatches to become a tadpole, which becomes a frog; deciduous trees lose their leaves annually and then grow new ones. The power of biological continuity and change over generations, however, is at least as impressive. It has long been harnessed by human breeders who have selected desired organisms to breed new kinds of roses or dogs and keep those breeds true. But it existed long before humans.

Using Energy

Life would not exist without energy. Defined by physicists as the ability to do work, energy pours onto Earth from the Sun, melting snow, raising winds, and heating mountains and plains. Ultimately, solar energy also supplies the power that drives most living things.

Certain organisms, however, living around hydrothermal vents in the sea floor, derive their energy not from the Sun but from chemicals in the superheated water. Yet even they demonstrate life's need for energy. From microbes to plants to animals, living things are perpetually transferring energy into themselves, using it, and gaining more.

Plants and certain plantlike organisms are able to use solar energy directly to make their own food. This process, which also involves the plants' use of inorganic chemicals such as water and carbon dioxide, is called photosynthesis. Most other organisms must get their energy either by eating plants or eating animals that have eaten plants.

REGULATION

Energy transfer takes place not only on the scale of individual organisms, but also on the smaller scale of the cells that make up their bodies, and on the larger scale of the ecosystems that comprise them. Similarly, regulation and interdependence take place on every biological scale, from cells to organisms to eco-systems. Regulation is necessary for energy to be useful, and interdependence is the condition living things must accept for having energy available.

Cells regulate themselves by maintaining control of their own biochemical processes. In organisms such as reptiles or mammals, control and coordination are maintained by the nervous system, including the brain, as well as the endocrine system, which produces hormones. Even the background state of living things tends to be maintained within a narrow range of conditions optimal for survival—for example, the stable body temperature and blood pressure of a human. For this stability, organisms rely on feedback mechanisms, such as the signals from heat receptors in the skin that tell the human brain to tell the sweat glands to start cooling an overheated body with perspiration.

Below: A collared lizard basks in sunlight, which helps it maintain a stable body temperature. Top left: Organisms such as these mussels, spider crabs, and worms derive energy from an underwater hydrocarbon seep.

This ability to remain internally stable, homeostasis, is an aspect of regulation that also occurs on a wider scale. The various activities of plants, animals, and other organisms in a region tend to form a dynamic balance that produces stability across an ecosystem.

INTERDEPENDENCE

Living things rarely exist in isolation. They usually depend on other organisms for many things. Plants depend on the wastes and decomposed bodies of animals and other organic material to provide nutrient minerals for use in photosynthesis. Flowering plants depend on bees, birds, and mammals to aid them in reproduction by spreading pollen or dispersing seeds. Herbivores depend on plants for food, and carnivores depend on herbivores for food. In many species, including all those that reproduce sexually, individuals depend on each other for such benefits as mating, parental care, cooperation in hunts, and maintenance of hives. Ecosystems are great networks of interdependence, in which the region's living things depend on each other in ways both obvious and unseen.

In some cases, interdependence microscopically. For example, the bacteria in our stomach assist our digestion even as we provide them with a home. Lichen, which appears to be a single primitive plant, is actually an organism consisting of an alga and a fungus living in a state of close association called symbiosis.

Top: These lichens growing on an alder tree on the Oregon coast are an example of symbiosis, a state of close association between individuals of different species. Each lichen consists of an alga, which makes food, and a fungus, which supplies water. Bottom: Bees depend on flowers for food, in the form of pollen and nectar. In turn, flowers depend on bees for aid in reproduction; bees spread pollen from one flower to another, fertilizing the plants they visit. Bees are also interdependent on each other, living in highly structured societies inside their hives.

Structures That Work

Roadside pebbles, Jupiter's hydrogen clouds, and other nonliving things have no function: there is nothing that they are for. Their structures can be explained simply by describing the physical processes through which they arose. But in a living thing, each component is typically directed toward some goal: it does have a function. The hand is for grasping; the leaf for making food out of sunlight; the exoskeleton for protection. This is because living things, unlike nonliving ones, evolved through a process of natural selection that favored the emergence of any structure that led to greater genetic success. In living things, there is almost always a close relationship between the structure of a thing and its function in furthering the survival of an organism's genes.

For this reason, biological structures cannot be explained simply by describing the processes through which they developed in the embryo. Instead, the biologist usually must explain what function the component serves, how the fulfillment of that function furthers the propagation of the organism's genes, and how this particular structure is adapted to that function.

Above: Octopus tentacles. Top left: An infant's hand grasping an adult's thumb. The appendages of both humans and the octopi are adapted to promote genetic survival.

STRUCTURE FOR USE

The relationship of structure to function can be found at many levels of biological organization, from organelles to organisms. The chloroplast, an organelle within plant cells, can carry out photosynthesis because it houses smaller sacs called grana that contain chlorophyll, a chemical molecule that can absorb solar energy. Neurons are long and branched so that these nerve cells can receive electrical impulses from other neurons and pass them on to others. The arms of an octopus are lined with suckers to seize prey and move along the sea floor. An earthworm's entire body is suited for life burrowing in the soil, from its lack of eyes to its ability to propel itself through muscular contractions.

Structure is not always perfectly related to function. The appendix, a tube that extends from the first part of the large intestine, has no known function in humans. It is believed that it once did have a function—aiding in digestion—but lost it in the course of evolution. Evolution generally succeeds in fitting structure to function, but not always perfectly.

BEYOND THE INDIVIDUAL

Indeed, in some cases, an organism's entire structural plan is not

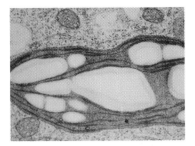

Chloroplast. This peanut-shaped structure within a plant cell has the function of producing food through photosynthesis.

A bat's wing has a double membrane of skin, not feathers, that makes flight possible.

OF BATS AND BIRDS

Sometimes species have similar structures because they are closely related. But in other cases, species that are not closely related develop similar structures just because that type of structure fits a certain function. An example can be found in the wings of bats, which are mammals, and birds, which are not.

When the ancestors of bats and birds independently took to the air, they needed wings to produce lift, an aerodynamic force that acts upward. A moving wing produces lift because it generates a greater pressure on the lower than the upper surface. To be effective, a wing also needs to be streamlined to minimize drag, the force that resists forward motion. Birds achieved this with feathers; bats with a double membrane of skin. But both bats and birds evolved wings that are aerodynamically functional because of their similar structures.

obviously related to the survival of its genes. Soldier termites have hard heads and powerful jaws well adapted to defending the termite colony, yet they do not reproduce. How does their soldierly architecture help them? It does not help them directly, but it helps the survival of the genes that they share in common with their close kin, the reproductive caste of the colony. Those genes, if transmitted, will lead to a new generation of soldier termites. The relationship of structure to function has to do with propagation of genes, not individual success in reproducing.

Structure operates with functions on still larger scales than organisms and social groups of organisms. In ecosystems at large, each organism has a function. Wolves control the population of elk, keeping them from becoming too numerous. Too many elk would upset the the available food supply and lead to their starvation. Trees provide habitat for birds and squirrels. Take away the organism —exterminate the wolves; fell the trees—and the ecosystem changes. Biological structures do not exist for the sake of ecological functions, but they perform key roles in maintaining ecosystems.

In termite colonies, only the reproductive caste passes on genes through reproduction, while the worker and soldier castes, which are their kin, support them.

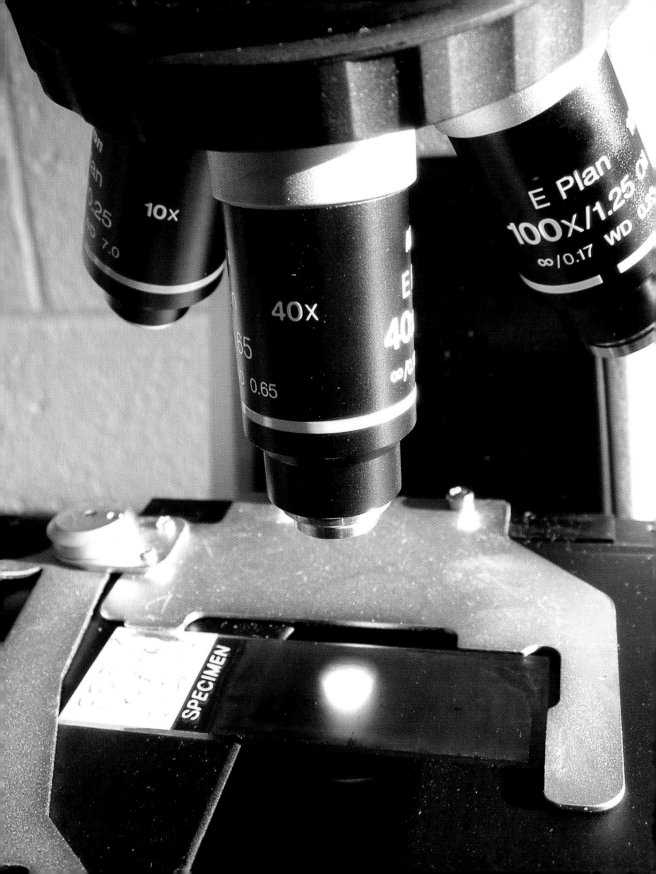

CHAPTER 2

FROM MICROSCOPES TO CAT SCANS

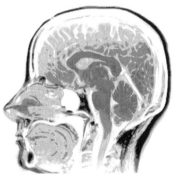

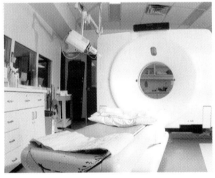

Life is too vast a subject for only one method of investigation. Biologists therefore approach it from many avenues and with many techniques, using tools that range from the microscope, first developed in the seventeenth century, to the CAT scanner, a late twentieth-century technology, to the latest supercomputers. Biologists work in laboratories, hospitals, nature preserves, and ocean depths. They belong to an ever-expanding variety of subdisciplines, each with a growing base of knowledge, which include botany, zoology, marine biology, microbiology, anatomy, physiology, genetics, mole-cular biology, and ecology. Yet, despite the diversity among biologists, all share certain fundamental practices. They all study living things, and they do so through scientific methods based on observation and experimentation.

Biologists also draw upon the insights of other disciplines, such as chemistry, physics, and computer science. And their work often produces social benefits. Medical advances have long depended on biological investigation into the causes of disease. Biological research into the breeding and care of crops and livestock has been important to agricultural growth. These are two of many examples of the power of biological research.

Top: CAT scan image. Bottom: CAT scanner. Left: Microscope. Biologists use many tools to investigate living things, from microscopes, which magnify objects, to computerized axial tomography, or CAT scans, an X-ray system used for precise body imaging. Microscopes date from the seventeenth century, while CAT scans date from the late twentieth century.

Doing Science

Biologists investigate life, but even more fundamentally biologists are scientists. As such, they share certain basic traits with their fellow scientists, whether they be astronomers, physicists, geologists, or chemists. The most essential characteristic that unites all scientists, no matter what they study—stars, atoms, rocks, chemicals, or organisms—is their reliance on observation and experimentation.

Scientists do science through close observation of nature. They may use sophisticated tools—the Hubble Space Telescope for astronomers or gene sequencers for biologists—but their primary goal is to "look" at nature and describe and interpret what they "see." A great step forward in biology came when Renaissance anatomist Andreas Vesalius began to report what he actually saw inside human cadavers—even if it contradicted what had been passed on for centuries from Greek physician Galen.

EXPERIMENTS WITH LIFE

Along with observation and description, many scientists make use of experimentation. In what is called the scientific method, the scientist generates a hypothesis, which is a tentative conclusion, and from that makes a prediction that can be tested by a controlled experiment. In such a trial, the scientist holds constant as many variables as possible so that one and only one variable at a time can be tested. The results of the experiment provide evidence either for or against the hypothesis—for it if the prediction is supported, against it if the prediction is not supported.

Not all biologists are engaged in generating hypotheses or testing them with experiments. For example, the scientists who spent years mapping the human genome, or human genetic inheritance, in the Human Genome Project were gathering data on genes, not experimenting with them. Some biological sciences do not lend themselves to experimentation: paleontologists, who study extinct life forms, cannot reanimate the dinosaurs to see if their theories about them are correct. But many biologists do conduct experiments as an important part of their work, such as those administering a particular microbe to a group of lab mice to test the hypothesis that the germ is the cause of a certain disease.

Above: A plant pathologist examines fungi cultures in a petri dish, probing for a potential use. Top left: Laboratory cultures of Phomopsis, *a destructive mold.*

credence, they must publish them in peer-reviewed journals and defend them when called upon at scientific conferences. They may need to collaborate with biologists from other specialties—a microbiologist with an ecologist, for example, for a study of how particular species of bacteria spread in tropical rain forests. Whether they work for a university, government agency, or corporation, they will need to obtain funding for their research. Like all scientists (and people in most other professions), biologists can be plagued by turf battles and career worries. But for the most part, biologists, like other scientists, are dedicated to the pursuit of knowledge that brought them to the job.

Left: A scientist at a microscope. Below: A microbiologist (left) and a chemist at DNA sequencer. Biologists engage in many kinds of tasks, from the scientist examining leaf litter for blacklegged tick nymphs to researchers using an automated DNA sequencer to obtain detailed genetic analysis of an unidentified microbe.

WORKING IN UNISON

Like all scientists, biologists do not work alone. From their predecessors, they inherit a body of knowledge that serves as their starting point, but that can be corrected in the light of new evidence. That body of knowledge includes established theories: large-scale frameworks for understanding, such as the theory of evolution and the cell theory, which have repeatedly been tested and confirmed.

In addition, biologists, like all scientists, work with their peers. For their findings to be given

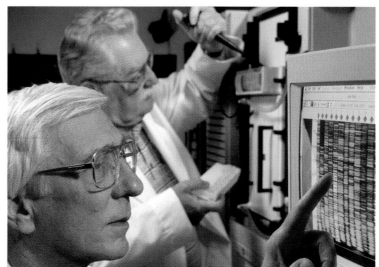

What Have Biologists Done for Us?

Biology may seem of limited importance to nonbiologists. When a new species of frog is found in Sri Lanka or a weed has its genome sequenced, some people might be forgiven for thinking that biology is of merely academic interest. But from its origins to the present day, the study of biology has been of vital importance to society.

The most obvious contributions have been in the field of medicine. The discoveries of biologists such as William Harvey, who discovered the circulation of the blood, laid the basis for the modern medical understanding of the human body. Every time a patient is examined or a treatment prescribed, the physician draws on biological knowledge.

The germ theory of disease, developed in the nineteenth century, formed the basis for such medical innovations as surgeons washing their hands before surgery, vaccinations for polio, and antibiotics for pneumonia. Medical advances today, such as the use of monoclonal antibodies in treating various diseases, still depend on biological discoveries.

AGRICULTURE AND ECOLOGY

Agriculture has also benefited from biology. The development of genetics made it possible to breed plants and animals scientifically, allowing for the development of new crops and livestock breeds. During the 1960s, in the Green Revolution, agricultural scientists introduced high-yield varieties of wheat and rice that helped poor countries such as India and Mexico increase their food supplies. More recently, ever more sophisticated methods have been applied to farming. Genetic engineering has been used to modify crops to make them resistant to insects, disease, or herbicides. This trend toward biotechnologically altered foods has aroused protest in some quarters, but is also widely viewed as a promising way to feed greater numbers of people.

Biologists, including ecologists, who study plants and animals in their natural environment have provided science-based counsel that helps society wisely manage natural resources. Marine biologists are a case in point. With their knowledge

Above: A microbiologist at the Centers for Disease Control and Prevention examines reconstructed influenza virus from the 1918 pandemic. Top left: Barley is one of the crops that is genetically engineered to resist viral infection.

about underwater ecosystems, they offer advice to governments making decisions about fishing, aquaculture, ocean dumping, pollution, and recreation.

OTHER APPLICATIONS

Some branches of biology may seem too arcane to be of much practical consequence.

Above: William Harvey. British physician Harvey's seventeenth-century discovery of the circulation of the blood contributed to the modern medical understanding of the human body.

Evolutionary biology, for example, might appear to be of interest only to those who care about trilobites and Neanderthals. But insights from evolutionary theory play an increasing role in many areas of life. Among them: coping with insect pests and bacteria that have evolved resistance to insecticides and antibiotics, assessing the gene pools of endangered species, and understanding the links between genes and human diseases.

Despite all the practical benefits of biology, the most important benefit is just in satisfying the age-old human thirst for knowledge. Aristotle, one of the earliest biologists, said "All men by nature desire to know," and that desire has not diminished in the two millennia since. Whether gaping at dinosaur skeletons in natural history museums or watching animals in the wild on a nature channel, people are fascinated by life, and biology helps us understand it.

A NEW KIND OF FINGERPRINT

In addition to providing many life-saving benefits, biologists have also developed forensic procedures that offer vital data to identify people. In 1977, British geneticist Sir Alec John Jeffreys (b. 1950) developed the

Above: A DNA sample with graph showing results of DNA analysis. DNA fingerprinting has become essential to forensic science. Top: A technician reading DNA sequences.

techniques for genetic or DNA fingerprinting, which uses variations in the genetic code to identify individuals uniquely. There are such genetic variations that it is extremely unlikely that any two unrelated individuals have identical DNA. For that reason, DNA fingerprinting has been used since the mid-1980s to identify criminals and resolve paternity disputes. It can also be applied to nonhuman species, for example in studies of wildlife population genetics.

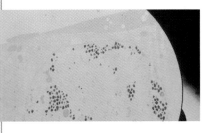

Getting the Job Done

In trying to understand living things, biologists, like all scientists, use tools and methods. But the tools and methods vary greatly according to the specific subdiscipline of biology.

Consider the field of marine biology. One marine biologist may descend to the ocean depths in a high-tech submersible to study deep-sea life. Another biologist, studying plankton, may just walk onto the dock with a bucket and collect surface water. Back in the laboratory, a centrifuge (a device for spinning samples rapidly) would be used to concentrate the plankton, after which the plankton would be grown in glass flasks containing seawater and nutrients. Still another marine biologist may use a hydrophone, or underwater microphone, attached to a buoy to listen for whale calls.

Ethologists, biologists who study animal behavior, may need camping equipment for long stays observing animals in the wilderness.

They may also use radio collars, or even satellite collars, to track the animals that they are studying. A herd of caribou in Alaska and Canada has been equipped with such satellite collars, which transmit the location of each animal to a satellite, which retransmits the data to Earth.

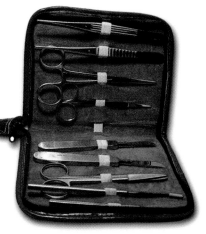

Dissection kit. For anatomical study, biologists may use such standard tools as scalpel, forceps, and scissors.

THE RESEARCHER'S TOOLS FOR MICROSCOPIC WORK

Biologists who study life on the microscopic scale, such as those in microbiology (the study of microorganisms) and cytology (the study of cells), depend on microscopes. All microscopes magnify objects, but they vary in power. Optical or light microscopes, which operate by refracting (bending) light through one or more lenses, can only magnify objects 2,000 times. For higher magnifications, biologists use electron microscopes, which employ a beam of

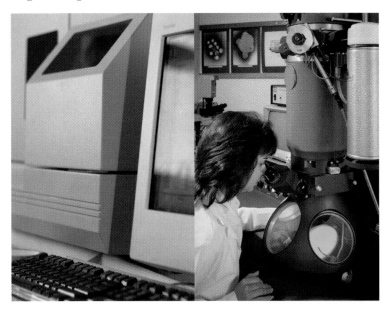

Far left: DNA sequencer. Left: Biologist at electron microscope. Tools for investigating small-scale phenomena as minute as viruses include the electron microscope and the DNA or gene sequencer. Top left: Cell nucleus infected by virus.

electrons instead of a beam of light to magnify objects. Electron microscopes, which can magnify objects more than 200,000 times, come in two principal varieties: the transmission electron microscope, which passes a broad beam of electrons through an ultra-thin specimen slice; and the scanning electron microscope, which sends a focused beam across an intact specimen's surface.

Studying the microscopic world involves more than just microscopes. Sometimes microbes must be cultured—grown in a petri dish containing a culture medium—so that they can be examined in greater quantity. Dyes may be used to color cell structures, making them more easily visible under the microscope.

EXPANDING THE TOOL KIT

Biologists studying anatomy may need the standard tools of dissection—scalpel, forceps, probe—but they may also need to make finer cuts. They may need laser microdissection tools, for example, to isolate single cells from a tumor for comparison to healthy cells.

Molecular biologists operate on even a smaller scale, studying the workings of molecules within living systems, particularly the DNA molecule that carries genes. Their tools include the gene sequencer, a tool for determining the order of nucleotides (the building blocks of DNA) or smaller chemical units in a piece of DNA. Another tool used in genetic engineering is the restriction enzyme. The dozens

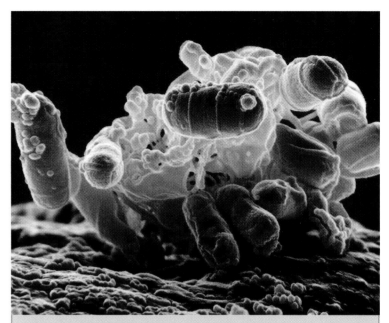

Electron micrograph of E.coli bacteria.

THE ELECTRON MICROSCOPE

Since the first electron microscope was developed by Ernst Ruska and other German scientists in 1931, it has come to be an important research tool in biology, chemistry, medicine, and metallurgy. In the mid-twentieth century, the electron microscope was instrumental in bringing about discoveries in cytology. Such discoveries relating to the structures of mitochrondria and ribosomes, important components of cells, were made by Belgian biologist Albert Claude and others. In medicine, transmission electron microscopes are useful for viewing very small structures (those having dimensions smaller than the wavelengths of visible light), such as viruses. The transmission electron microscope magnifies a specimen by up to 1,000,000 times. The other type, the scanning electron microscope, can magnify a specimen by up to 500,000 times.

of restriction enzymes that are known cut molecules of DNA at particular sites, generating fragments of DNA.

Sometimes a device developed for one subfield of biology has applications in another field. Computerized axial tomography, or CAT scans, an X-ray system used for precise body imaging, has gained wide use in medicine, but it also help paleontologists

to study fossils. Paleontologist Larry Witmer has used the technology to scan dinosaur skulls.

Biologists use many other tools, from manometers (that measure the uptake of gases in photosynthesis and respiration) to supercomputers, to fill in the blanks in models of DNA. As technology continues to improve and expand, biologists are certain to take advantage of it.

In the Field

Some biologists are primarily concerned with how an organism lives in its natural environment. These are sometimes called organismic biologists, to distinguish them from cellular biologists, who are more concerned with what happens at the cellular or molecular level within an organism.

But the main difference between them may be that organismic biologists, to get a good look at their species, are often out in the field rather than the lab. Hence, many of them are better known as field biologists.

Studying living things in their natural habitats, field biologists include botanists (who study plants), zoologists (who study animals), marine biologists (who study ocean life), ethologists (who study animal behavior), ecologists (who study the relationships of living things to each other and the environment), and wildlife and fisheries biologists (who study how to conserve stocks of wild fauna). Biologists often specialize even further: one zoologist may belong to the branch called herpetology, the study of reptiles and amphibians; another to entomology, the study of insects. In many cases the specialization is even finer: for example, one entomologist may

Measuring soil quality. Agricultural Research Service scientists work as a team, checking a soil sample (left), drawing another (right), and operating a GPS device (center).

be a forest entomologist, specializing in timber pests, while another is a lepidopterist, specializing in butterflies and moths.

WORKING IN THE FIELD
To study organisms in their natural habitats, field biologists often must rough it in difficult conditions. They may spend long periods on chilly Arctic ice, following polar bears; in hot southwestern deserts, tracking

Left: A marine biologist gets a firsthand view of ocean life in the Wasp, an articulated diving suit with propellers that make it fly through the water. The Wasp's artificial limbs, or manipulators, are operated from the inside. Top left: Marine biologist studying life on a coral reef.

digital cameras and digital pens/notebooks; and even old-fashioned paper notebooks. They need to analyze what they have found and share it with colleagues.

IMPACT OF FIELD BIOLOGISTS

Despite the remoteness of the places to which field biologists may go, their research often has a profound effect on biology and other areas of life. Charles Darwin's theory of evolution owes much to his 1831–36 expedition aboard the HMS *Beagle*, during which he studied many plants and animals. The highly unusual specimens he found on the Galapagos Islands off the coast of Ecuador helped him formulate his theory of natural selection.

Although some field biologists are affiliated with universities, many are employed in non-academic venues. Government agencies hire field biologists to help manage wildlife, control pollution, or enforce environmental laws. Private corporations find numerous uses for field biologists, such as monitoring and managing the cooling lakes around generating stations. Some field biologists go into business for themselves as independent consultants advising government and industry about issues such as environmental landscaping and habitat improvement.

In an agricultural application of biology, a scientist measures severe soil erosion in a Washington State wheat field.

kangaroo rats; or in sultry rain forests, studying tropical vegetation. Some areas of the world become particularly well studied. Barro Colorado Island, an island of six square miles in Panama's Canal Zone, has been home to field biologists since the early 1900s, making it one of the best-studied pieces of rain forest in the world.

The primary task for field biologists is observing: watching what plants and animals do in the wild. They may need to collect samples and tag animals for monitoring. They need to capture data with an array of tools—wireless sensor networks to obtain temperature and other environmental data;

Jane Goodall with chimpanzee.

THE GOMBE CHIMPANZEES AND JANE GOODALL

In 1960, the British zoologist Jane Goodall arrived at Tanzania's Kakombe Valley to begin an in-depth study of the wild Gombe chimpanzees. The project she thought would take months went on for years, and reversed assumptions about chimpanzee behavior and the nature of field biology.

Before Goodall's studies, chimpanzees were thought to eat fruits and vegetables, and occasionally insects and rodents. Goodall observed that they also hunt larger animals; she was the first to observe a group of chimpanzees killing another group for nonsurvival reasons. She also watched them refashion twigs into hunting tools. These observations showed that warfare and weapons-building were not activities limited to humans.

Goodall's discoveries could be made only after an extended period of close observation. Her pioneering work made extended close-range study accepted practice for field biologists.

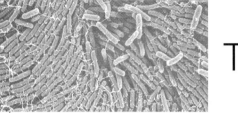

The Microscopic World

Much of the living world is not visible to the naked eye. With the unaided eye, no one can see the cells in one's own hand, or the bacteria on the bathroom sink, or the strands of DNA that is the blueprint for the resemblance between parent and child. Yet, for some biologists this invisible world is the stuff of daily life. Using optical and electron microscopes, bacterial and cell cultures, restriction enzymes, gene sequencers, and other laboratory technology, they penetrate the microscopic world. Some of these scientists are microbiologists, who study microscopic organisms (microbes) such as bacteria, viruses, protozoans, algae, molds, and yeasts. Often they specialize in certain types of microbes. For example, bacteriologists concentrate on bacteria, virologists viruses, and mycologists fungi. Some microbiologists focus on one aspect or another of the interaction between microbes and humans, other organisms, and the environment. Medical microbiologists study how microorganisms cause diseases and look for treatments or cures. Agricultural microbiologists investigate such issues as plant diseases and the role of microbes in soil fertility and spoilage.

CELLS AND MOLECULES

Other biologists who study the microscopic world include cytologists, who investigate the structure and function of cells. Their microscopic examinations of cells are important in helping to understand and diagnose diseases such as cancer. They often work with other disciplines, creating hybrid fields such as cytochemistry, the study of chemical activities inside cells, and cytogenetics, the study of the behavior of chromosomes and genes. Cytology can be considered a branch of histology, the microscopic study of tissues.

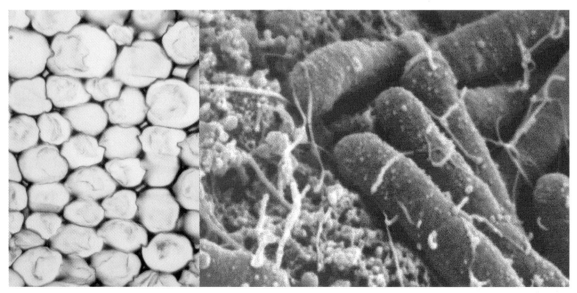

Above left: A cluster of cells. Above right: Bacteria attached to a surface. Top left: Bacteria. All three images were taken using an electron microscope. Microbiologists employ equipment that allows them to see many sights that are invisible to the naked eye. Among these: bacteria aligning side to side as they accumulate on surfaces and anchoring themselves to surfaces with extra-cellular polymers.

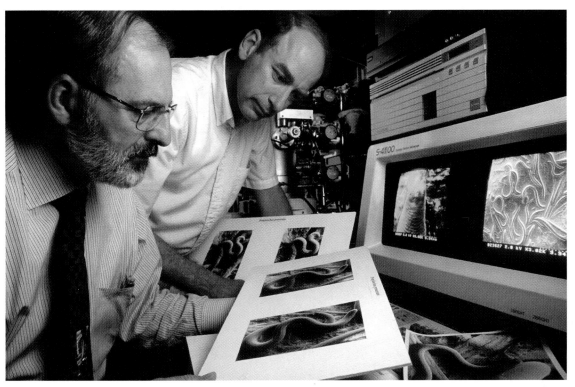

A cytologist (left) and zoologist study the nematode Caenorhabditis elegans. *It is used as a model for studying the biochemistry of microscopic plant parasites to which it is related, soybean cyst and root-knot nematodes, both of which prey on crops.*

Molecular biologists study molecules within organisms, especially the DNA molecule that carries genes. They have been behind many of the breakthrough products of biotechnology, a set of techniques that applies biological processes to the production of materials for medicine and industry. For example, the first commercial product that resulted from genetic engineering, marketed in 1982, was human insulin for diabetes patients. The insulin was produced by a type of bacteria that had been genetically altered to produce it. This was due to molecular biological research that isolated the human genes that code for insulin, and to genetic engineering that enabled the insertion of the genes into the DNA of the bacteria.

GENOMES AND EMBRYOS

Genomics researchers study the entire set of genes carried in each cell of an organism. This burgeoning field includes several subcategories: structural genomics, the mapping of the complete genome of an organism, as occurred in the Human Genome Project; functional genomics, the study of how genes are expressed and how their products work; and comparative genomics, which searches for regions of similarity between the genomes of different species, such as a worm and a human. Knowing what a DNA sequence does in one species may shed light on what a similar region does in another, and may also help clarify the evolutionary relationship of the two species.

Other biologists enter the microscopic world only when the area of study fits their research. A marine biologist studying swordfish might not have much use for a microscope, but he or she might when studying plankton, the tiny organisms that live at or near the water surface. Similarly, embryologists (biologists who study the development of organisms) would need a microscope when a specimen is a fertilized egg cell, but not when it is about to be born.

The Many Sides of Life

The array of specialties for biologists is as broad as life itself. For example, some are experts on anatomy, the study of the structure of organisms. The arrangement of bones, the shapes of organs, and the branching of trees are typical concerns. Some are comparative anatomists, who compare the structure of different organisms, such as apes and humans, often with a view to determining evolutionary relationship. The family resemblance between apes and humans is so close

it extends to the pattern of ridges on each species molars. Some biologists must deduce the structure of organisms from fossil remains: these are the paleontologists, who study the organisms of past geological ages.

Some biologists, known as physiologists, focus not on the structure of organisms but on how they function. Their work often has medical connections. A cell physiologist, for example, studies the functioning of individual cells, perhaps with the aim of learning why some cells suddenly turn cancerous. But physiologists may also do basic research on a range of organisms. When

scientists first investigated how a camel can survive so long without water, that was physiology; when today they study the workings of a newly discovered tropical flower, that is also physiology. Comparative physiologists compare the functioning of organisms; like comparative anatomy, this research can illuminate evolutionary relationships.

GENETICS AND SYSTEMATICS

Geneticists study heredity, coming at the topic from different angles (or disciplines). Population genetics studies the processes that change the frequency of genes over time in a population of organisms. Behavioral genetics studies the relative importance of genes and environment in determining how an animal behaves. Transmission genetics applies and extends Gregor Mendel's laws of genetics to study patterns of inheritance. Molecular genetics studies the molecular basis of genes, overlapping with molecular biology, the study of molecules within organisms.

Systematics, also called taxonomy, is the study of how to classify organisms according to their natural relationships. This field also branches into subfields. Molecular systematics uses the sequence of amino acids

Left: Anatomical drawing of orangutan and human skeletons. Comparative anatomy is one of the biological subfields. The other is bioengineering, which is responsible for producing the artificial hip. Top Left: An artificial hip undergoes testing.

or nucleotides (building blocks of proteins and nucleic acid, respectively) to determine the evolutionary relationship of organisms. Biosystematics combines all, classifying the diversity of species through genetic, biochemical, and other observational studies. The resulting data yields important information on which other fields of biological research depend.

REACHING OUTSIDE BIOLOGY

Biologists frequently draw on other scientific disciplines to advance their research. Biochemistry uses chemistry to study the chemical processes within living things. By tracing the chemical pathways through which organisms metabolize food or fight infection, biochemistry has contributed to many other fields of study, including physiology and medicine. Psychologists, who study behavioral and mental processes, combine the biologist's attention to organisms with the social scientist's concern for social settings.

Biophysicists apply physical laws and techniques to the study of life. Bioinformatics uses computerized systems to collect, store, and analyze data from DNA and protein sequencing. Biomechanics studies the structure and function of living things using the methods of mechanics, the study of the effects of forces on matter. Biomechanics has applications for everything from preventing knee injuries in football players to understanding the fluid mechanics of a swimming fish. Bioengineers go further, using biomechanical knowledge to develop artificial hips, heart valves, and other replacement parts.

A systematist (foreground) and technician use an automated gene sequencer to investigate the relationships between two organisms, Fergusonina *flies and* Fergusobia *nematodes. These organisms are known to have a symbiotic relationship.*

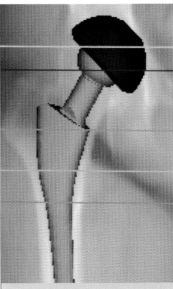

Artificial hip.

THE ARTIFICIAL HIP

More than 168,000 hip replacement surgeries are performed yearly, and a great deal of science goes into each one. To develop a workable replacement for the ball-and-socket joint that makes up a human hip requires the integration of biomechanics and bioengineering. The two-part prosthesis, or artificial hip joint, that replaces the diseased hip joint is made of two components: a metal ball component and a plastic socket component (which may have a metal outer shell). The prosthesis may be attached to the bone with a type of surgical cement or with a fine mesh, so that the bone can grow into the mesh and attach naturally. In either case, the result is a replacement hip that can be integrated into the human body and function like a human hip.

CHAPTER 3

CENTURIES
OF SCRUTINY

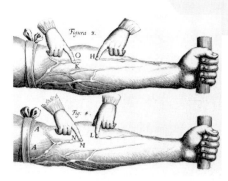

Left: French chemist Louis Pasteur at work in his laboratory. Pasteur's research helped develop germ theory in the nineteenth century. Top: Illustration from British physician William Harvey's book on the circulation of the blood. Bottom: Illustration from Das Meer, *a book on ocean life by German botanist Matthias Schleiden. Over the ages, biology has advanced by steps small and large. The great leaps forward in the science of life have included Pasteur's development of the germ theory of disease, Harvey's discovery of how blood circulates in mammals, and Schleiden's collaboration with German physiologist Theodor Schwann on the cell theory.*

Scrutiny of the surrounding world is ancient; on cave walls prehistoric humans painted detailed images of the animals they hunted and later gained enough biological knowledge to domesticate many plants and animals. Descriptions by Aristotle of many species and Galen's writings on human anatomy became the long-standing references.

Despite its roots in time, modern biology begins with the Scientific Revolution of the sixteenth to eighteenth centuries, when scientific methods were first applied to the age-old questions about what life was and how it could exist. In the sixteenth century the work of Andreas Vesalius exploded anatomical errors that had gone unchallenged for centuries. In the eighteenth century, Linnaeus developed a new way of classifying organisms, and in the next, Darwin explained that all organisms are related through evolution, Mendel showed that genes are the unit of inheritance, and Pasteur showed that germs cause disease. The twentieth century saw a further explosion of knowledge, as medicine raced forward and Watson and Crick revealed the structure of DNA. The early years of the twenty-first century have already witnessed the completion of the Human Genome Project, and no doubt further biological advances are ahead.

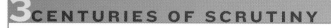

Biologists of the Ancient World

As humans evolved, they depended for survival on many other organisms: the plants whose roots and nuts they gathered for food; the fish and crustaceans they caught; the deer and mammoths they hunted. Many of their tools were made from wood or bone; their dyes, poisons, and medicinal herbs were products of wild bushes and flowers. Their dependence on nature trained them to observe the organisms around them, distinguish and classify them, and experiment with their uses. This was the beginning of biology.

Starting about 30,000 years ago in France and Spain, prehistoric human beings called Cro-Magnons demonstrated close observation of nature in detailed cave paintings of bears, mammoths, horses, bison, cattle, and other animals. In time, they applied their biological knowledge to the task of domesticating animals and plants. Around 14,000 years ago, dogs were domesticated from the wolf in Mesopotamia (now Iraq). Domestication of herd animals started with goats in Persia (Iran) about 12,000 years ago.

Agriculture began in several regions around the same time, probably independently: in northern Mesopotamia about 10,000 years ago, with wheat and barley; in Mexico about 9,000 years ago; and in China and Japan about 8,000 years ago.

THE GREEKS

While the biological research of prehistoric times had been decidedly practical, the advent of historic times (after the invention of writing) introduced biological scholarship, particularly in Greece. Alcmaeon, a Greek physician of the fifth century BCE, studied human bodies by dissecting them. Other Greek philosophers of the fifth and sixth centuries BCE, such as Xenophanes and Democritus, studied plants and animals, with Democritus performing comparative anatomical observations on many kinds of animals.

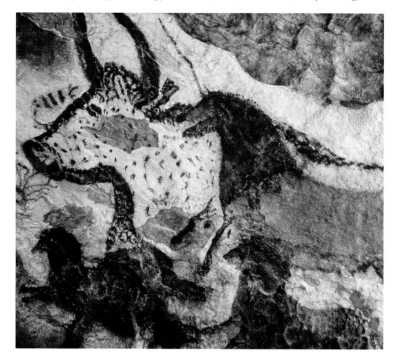

Left: Cro-Magnon cave painting from Lascaux, in the Dordogne valley of southwest France. Many ancient people were keen observers of the living world. Prehistoric humans known as Cro-Magnons recorded their observations of the animals they hunted on the walls of their cave dwellings. Egyptians in the New Kingdom, which ended about 1087 BCE, documented their long medical traditions in the Edwin Smith Surgical Papyrus. Top: The Edwin Smith Papyrus.

Greek philosopher Aristotle (384–322 BCE) brought together an immense body of biological knowledge. Unlike some philosophers of his day, he regarded sense experience as trustworthy, making him a forerunner of the modern scientific empiricist, who believes that sense experience is the most reliable basis for knowledge. He studied various aspects of animal life, including classification, parts, movement, and generation. He also investigated such matters as memory, sleep, dreams, youth, old age, respiration, and plants. Aristotle's pupil Theophrastus became known as the father of botany for delving even further in the study of plants, describing more than 500 plant species.

MEDICINE

Ancient physicians contributed greatly to the growing body of biological knowledge. Both the Chinese and Egyptians established medical traditions that are documented in writing, such as the manuscript on medicinal plants attributed to Chinese emperor Shen Nung (2800 BCE) and

Colericus (yellow bile), left, and Sanguineus (blood), right, two of the four humors.

THE THEORY OF THE FOUR HUMORS

One of the most durable biological theories handed down from antiquity was that of the four bodily "humors" (fluids). In the fourth to third century BCE, Hippocrates suggested that disease arose from an internal imbalance of one of the body's four humors: blood, phlegm, black bile, and yellow bile. This theory became accepted doctrine and over the centuries informed various types of medical treatment. One was bloodletting, the process of opening a vein to release any disease. The practice was common medical procedure from the Middle Ages into the eighteenth century. The advent of modern biology and medicine brought an end to the disease relevance of the four humors.

the Edwin Smith Surgical Papyrus from Egypt's New Kingdom (ending about 1087 BCE).

Greek physicians also made important contributions. Hippocrates (c. 460–c. 370 BCE) is considered the father of medicine, and not only for his famous "Hippocratic oath," which defined standard ethical practice for physicians. He is esteemed as a medical scientist for teaching that diseases had natural

In the fourth century BCE, Greek philosopher Aristotle penetrated many biological fields that are now an integral part of the modern biology canon.

causes and stressing bedside observation. Galen (c. 130–c. 200), a Greek physician in Rome, dissected animals and made numerous anatomical discoveries.

The teachings of Galen and Aristotle were impressive, but contained many errors. Nevertheless, they came to be regarded as authoritative in Europe during the Middle Ages. The practice of checking assumptions against observation of nature fell by the wayside, and advances in biology grew few. But in the sixteenth century, the situation began to change.

The Scientific Revolution

In the sixteenth century in Europe, as the Middle Ages ended and the period of cultural rebirth known as the Renaissance reached its peak, science entered a new age. The scholars who studied nature in all its forms—living things, chemicals, rocks, stars—realized that empirical observation, sense experience, was the most reliable basis for learning more about their subjects. They began to emphasize experimentation and quantification to increase precision, test hypotheses, and make discoveries. This period, from about 1550 to 1700, is known as the Scientific Revolution.

The spirit of the coming Scientific Revolution was already present in the anatomical drawings of Italian artist Leonardo da Vinci (1452–1519), who based his studies on dissections of human cadavers. Flemish anatomist Andreas Vesalius (1514–64), often regarded as the founder of human anatomy, also dissected cadavers, and from those studies published a systematic treatise, *On the Structure of the Human Body* (1543). He exploded many of Galen's errors, thus beginning to free biology from unwarranted reverence for ancient authorities.

British physician William Harvey (1578–1657) discovered how blood flows in mammals.

In 1628, he demonstrated that the heart is a muscle, which operates like a pump, driving the blood through regular contractions.

THE MICROSCOPE

The development of the microscope was a great aid to biologists in the Scientific Revolution. Italian anatomist Marcello Malpighi (1628–94) used a microscope in 1660 to discover blood circulation through the capillaries. English scientist Robert Hooke (1635–1703) discovered plant cells, drawings of which he published in his book *Micrographia* (1665). He also originated the term "cell" because each, to him,

Above: Flemish anatomist Andreas Vesalius. Top left: Dutch scientist Anton van Leeuwenhoek. The Scientific Revolution of the sixteenth to eighteenth centuries, with its emphasis on empirical observation, brought many biological advances, including the anatomical discoveries of Vesalius and the microscopic findings of van Leeuwenhoek.

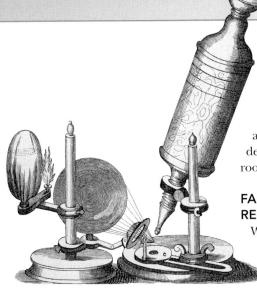

Microscope of English scientist Robert Hooke, discoverer of plant cells.

correctly. He studied animals and plants on the microscopic level, showing, for example, how sap moves through a network of channels and describing the structures of roots, stems, and leaves.

FAR-RANGING RESEARCH

With or even without microscopes, research by biologists in the sixteenth century seemed limitless. Flemish physician Jan Baptista van Helmont (1579–1644) studied the metabolism of the willow tree. In another study on fermentation, he coined the word "gas" for the peculiar air that arose from the process, now known to be carbon dioxide.

Also in the sixteenth century, Italian scientist Francesco Redi (1626–97) used ingenious insect experiments to disprove spontaneous generation, the theory that life can originate from nothing. French philosopher René Descartes (1596–1650), who studied animal physiology as well as philosophy and mathematics, introduced the concept of reflex action. Danish anatomist Nicolaus Steno (1638–86) coined the term "ovary" after discovering ovaries in a shark. Dutch naturalist Jan Swammerdam (1637–80) studied the metamorphosis of insects. In 1694, German botanist Rudolph Jakob Camerarius (1665–1721) demonstrated that plants reproduce sexually.

It was the age when modern science was born, and yet it was only the beginning. Much remained to be learned, especially about the underlying principles that explained the dazzling variety of biological phenomena. Those principles would soon begin to emerge.

resembled an empty room.

Using microscopes, Dutch amateur scientist Anton van Leeuwenhoek (1632–1723) discovered spermatozoa and made descriptions of bacteria, protozoa, and other microorganisms. In 1674, he became the first to describe red blood cells

SPONTANEOUS GENERATION

Although to this day it may seem that unwanted life forms (pests or weeds, for example) arise spontaneously, Italian physician Francesco Redi proved otherwise. In 1668, he set up an experiment to discover whether maggots originate from rancid meat. He placed various meats in eight containers and covered four of them. Over time, all of the meats turned rancid, but only the uncovered vessels had developed maggots.

To eliminate the chance that an absence of air kept maggots away, Redi conducted repeated experiments. This time, he covered four of the meat-containing vessels with gauze. Again, the four uncovered containers had maggots; the covered ones did not. Redi concluded that the maggots did not arise spontaneously but from another form of life. Flies laid eggs on the meat and those eggs hatched into maggots.

Through a series of insect experiments, Italian physician Francesco Redi disproved spontaneous generation, the view that life can originate from nothing.

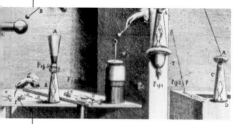

A New View of Life

In both the eighteenth and nineteenth centuries, biology took great steps forward in the understanding of life. Behind the progress was, in part, a surge in the number of living things to study, as European explorers penetrated distant regions and brought back specimens and descriptions. Naturalists realized that they needed a system to classify new species as they were discovered, and Swedish naturalist Carolus Linnaeus (1707–78) rose

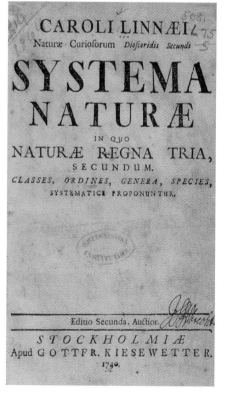

to the challenge. An authority on plants, he established the modern scientific system for classifying organisms and naming species. It included a two-part Latin name for every living thing: the first part for the genus, or group; the second for the species, or kind.

French naturalist Georges Cuvier (1769–1832) also made an important contribution to the science of classification, or taxonomy, especially with regard to animals. While comparing animal body types to classify them, he founded the modern field of comparative anatomy. He also helped found the field of paleontology by including the fossilized remains of prehistoric animals in his comparisons.

PHYSIOLOGY

Another field that underwent dramatic changes in the late eighteenth century was physiology. Until then, it had been widely believed that the functioning of living things depended on immaterial or spiritual forces different from those that activated nonliving

Left: Title page of Carolus Linnaeus's Systema Naturae. *Top left: Luigi Galvani's frog experiments.*

things. But by the late eighteenth century, many physiologists were building evidence for a mechanistic or materialist view of life, one that regards life as fundamentally physical and chemical.

Italian physician and anatomist Luigi Galvani (1737–98) discovered that nerve impulses were electrical in nature when he showed that electric currents could cause contractions in the muscles and nerves of dead frogs. This research formed the basis for neurophysiology, the physiological study of the nervous system. French chemist Antoine Lavoisier (1743–94), regarded as the founder of modern chemistry, applied chemical techniques to physiology, showing that respiration is like the burning of a candle in that both consume oxygen and generate heat and carbon dioxide.

THE ORIGIN OF SPECIES

For many eighteenth-century people, the mechanistic idea that life was entirely material, with no functional role for spirit, was an affront to religious faith. The clash between biology and religion heated up as biologists turned to the question of the origin of species. From Cuvier on, the findings of paleontologists raised questions about the connections, if any, between extinct species and living ones. Interest grew as dinosaur fossils began to be discovered and described, starting with *Megalosaurus* in England in 1824. British paleontologist Mary Anning (1799–1847) unearthed fossils of a variety of extinct species, including the ichthyosaur, pterodactyl, and plesiosaur. Did these beasts have any connection to modern life forms?

French scientist Jean Baptiste Lamarck (1744–1829), who coined the word "biology" by putting together the Greek words for life, *bios*, and study, *logos*, had proposed one answer. He had argued that organisms evolved by inheriting the acquired characteristics of their parents. But British naturalist Charles Darwin (1809–82) proposed a different view. In his landmark treatise *The Origin of Species* (1859), Darwin argued that evolution occurred through natural selection: the transmission not of acquired traits but of traits the parents themselves had inherited and that had made them successful in the competition to reproduce.

The Origin of Species has generated controversy from that day to this, but Darwin's theory of evolution eventually became a part of the consensus of biologists, indispensable to almost all their fields of study.

British naturalist Charles Darwin revolutionized biology with his theory that species originated through a process of evolution by natural selection. Although controversial when it was first published, the theory eventually became part of the consensus of biologists.

A cartoon caricaturing Charles Darwin. Because Darwin theorized that humans and simians had a common ancestor, he was frequently portrayed in cartoons as an ape or monkey. In this 1878 cartoon, published in France, he and French philosophical writer Emile Littré are both portrayed as monkeys performing a circus act.

The Basics of Life

In the nineteenth century, scientists advanced their basic knowledge of biology, including the structure, functioning, and chemistry of organisms. Early in the century, chemicals were categorized as either organic or inorganic, and it was widely believed that only living tissue could produce organic chemicals. But in 1828, German chemist Friedrich Wohler (1800–82) became the first person to make an organic substance from inorganic ones. The substance was urea, the principal waste product found in mammal urine. Wohler, however, produced urea by heating the inorganic substance ammonium cyanate. This discovery helped confirm the growing view that living things are composed of the same fundamental substances as nonliving things, and function according to the same physical laws. Wohler also staked a claim to being the founder of biochemistry, a field that emerged during this century.

By 1860, French chemist Pierre Berthelot (1827–1907) had synthesized a variety of organic molecules from inorganic substances, including methyl alcohol, ethyl alcohol, methane, and benzene. Organic chemistry came to be understood afterward as the chemistry of carbon compounds, rather than compounds that occur only in living things.

THE WORKINGS OF THE BODY

Even as understanding of biochemistry increased, so did knowledge of physiology—the study of how living things functioned. In the mid-nineteenth century, French physiologist Claude Bernard (1813–78), founder of modern experimental physiology, showed that control of body temperature in warm-blooded animals resides in the nervous system.

Left: French physiologist Claude Bernard. Top right: German pathologist Rudolf Virchow. Nineteenth-century scientists such as these made major advances in basic knowledge of biology, including the structure and functioning of organisms. Top left: An illustration of medusas, a type of marine organism, from Matthias Schleiden's book on ocean life.

He also developed the concept of homeostasis, or internal stability, as he demonstrated that animals could maintain constant body temperature and blood glucose concentration in spite of varying external conditions.

German physicist Hermann von Helmholtz (1821–94), renowned in physics for helping to establish the law of conservation of energy, applied physical methods to the study of the nervous system. With these techniques, in 1852, he determined the speed at which messages travel along nerves.

Knowledge of plant functioning also grew in this period. In 1840, German chemist Justus von Liebig (1803–73) showed that plants use carbon dioxide in the air to synthesize organic compounds but take nitrogenous compounds from the soil.

CELL THEORY

During the nineteenth century, scientists also developed the modern conception of the basic unit of organisms, the cell. In 1831, Scottish naturalist Robert Brown (1773–1858) recognized the nucleus as a regular structure in all plant cells. In 1834, French chemist Anselme Payen (1795–1891) discovered cellulose, the main constituent of the cell walls of plants.

In the 1830s, two Germans, botanist Matthias Schleiden (1804–81) and physiologist Theodor Schwann (1810–82), proposed the cell theory. The theory stated that the cell is the basic structural and functional unit of life. This was to become the accepted view among biologists and formed the basis for many other discoveries.

The cell theory was augmented by the work of German pathologist Rudolf Virchow (1821–1902), who in 1858 argued that disease results from disturbances in cell function. He was a founder of pathology, the study of diseased tissue. German zoologist Robert Remak (1815–65) also advanced the cell theory when he argued, in 1852, that cells originate in the division of other cells.

Jacket of Das Meer (The Sea), *German botanist Matthias Schleiden's 1869 richly illustrated book on marine animals and plants. With German physiologist Theodor Schwann, Schleiden proposed the cell theory, the idea that the cell is the basic structural and functional unit of life. This was to become the accepted view among biologists.*

Medicine Reborn

Medicine—the art and science of diagnosing, treating, and preventing disease—is deeply connected to biology. For that reason, the rapid changes in biology in the nineteenth century helped to transform medicine, making it more systematic and effective. In turn, evidence from medical practice helped to drive biological discoveries, most notably with respect to the growing view that germs are the cause of many diseases.

In 1847, Hungarian physician Ignaz P. Semmelweis (1818–1865) showed that puerperal sepsis, or childbed fever, could be prevented by having doctors wash their hands in a solution of calcium chloride before delivering babies. This was evidence that the doctors had been spreading the disease from patient to patient, and that chemical agents could stop the spread.

THE GERM THEORY

French chemist Louis Pasteur (1822–1895) investigated what was causing such contagious diseases. In the 1860s, he helped develop the germ theory of disease, which held that contagious diseases are caused by micro-organisms. By then, he was already an expert on microbes, having shown that they spoil wine, milk, and beer. To preserve the beverages, he introduced the use of heat to kill germs, the now-familiar technique of pasteurization.

Pasteur went on to identify the germs that cause various diseases and develop a method of prevention through vaccines: compounds containing weak or killed microbes that cause the body to become immune without suffering the disease. In the eighteenth century, British physician Edward Jenner (1749–1823) had already developed a vaccine for smallpox, but Pasteur was now able to explain how it worked and extend the principle to the creation of other vaccines. He invented vaccines for anthrax and rabies and laid the groundwork for the development of later vaccines.

German physician Robert Koch (1843–1910) carried on Pasteur's work. In 1882, Koch discovered the bacterium that causes tuberculosis. As his work helped advance the emerging science of bacteriology, Koch showed that specific bacteria are the cause of particular diseases.

He introduced new techniques for culturing bacteria and established Koch's postulates, which is the criteria to prove that a certain microbe is responsible for a given disease.

Left: Nineteeth-century caricature of French chemist Louis Pasteur. Top left: Microbes. Medical science grew rapidly in the nineteenth century, in part as a result of Pasteur's work on the germ theory of disease. Physicians learned that the unseen world of microorganisms was responsible for much of the illness and suffering they were trying to combat.

ADVANCES IN SURGERY

As physicians learned that germs are the cause of contagious disease, interest grew in the observations of Semmelweis that chemical agents could stop their spread. British surgeon Joseph Lister (1827–1912) was concerned about the high rate of mortality from surgery, and believed it might be caused by microbial infections. In 1865, he began applying a solution of carbolic acid (phenol) to wounds, instruments, dressings, and the surgeon's hands. The germicide reduced surgical mortality by more than half.

The technique proved successful. It revolutionized surgery, turning it from a desperate, often futile, measure into one with a reasonable chance of success.

Other techniques developed in the nineteenth century also made surgery more effective. The discovery of anesthetics, such as ether and chloroform in the 1840s, allowed patients to endure surgery painlessly. And at the turn of the century, in 1900, Austrian-born American pathologist Karl Landsteiner (1868–1943) discovered the major types of human blood, O, A, B, and AB. He demonstrated how knowledge of blood types could be used to carry out safe blood transfusions—a discovery that has been beneficial to surgery and other medical situations ever since.

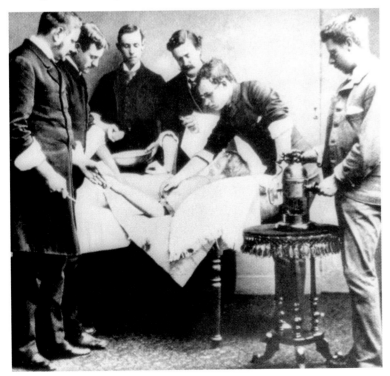

Surgery in 1883. By the late nineteenth century, antiseptic methods in surgery were widely used, largely due to the influence of British surgeon Joseph Lister. Beginning in 1865, he sterilized wounds, instruments, dressings, and surgeon's hands with a solution of carbolic acid (phenol). On the right is a device used to spray carbolic acid in the air.

In 1853, Queen Victoria accepted anesthesia for the birth of her eighth son, Leopold.

QUEEN VICTORIA'S DOCTOR

In 1853, Royal Doctor John Snow (1813–1858) persuaded his patient Queen Victoria of Great Britain to ignore taboo and tradition and accept anesthesia for the birth of her eighth child, Leopold. At that time, many physicians and clergymen opposed anesthesia, particularly for childbirth, because they considered it immoral to dull pain. By consenting to be anesthetized, Queen Victoria changed popular attitude toward anesthetics and Snow changed the face of modern medicine.

Before childbirth, Snow administered a colorless, volatile, sweet-smelling liquid known as trichloromethane, or chloroform. It was inhaled by the queen, and proved an effective childbirth anesthetic that did not harm the child.

By ingesting chloroform, Queen Victoria validated the use of anesthesia for childbirth and made it socially acceptable. In later medical research, chloroform was found to cause liver damage and was replaced by other sedatives. But the pain once considered divinely required became unnecessary.

Unlocking the Gene

Darwin's theory of evolution depended on the existence of a mechanism of inheritance, yet Darwin himself knew little about how that mechanism worked. The enigma was solved in the nineteenth and twentieth centuries, bringing not only progress in understanding evolution but practical outcomes in medicine, agriculture, and industry.

Unbeknownst to Darwin, a contemporary, Austrian botanist Gregor Mendel (1822–84) could have aided his inquiries. Mendel discovered that the inherited traits of organisms are transmitted by hereditary units (later called genes). Experimenting with the breeding of garden peas, he formulated the basic laws of heredity. His results were published in 1866 but went largely unnoticed by the scientific community until 1900.

By that year, three botanists, Dutchman Hugo Marie de Vries (1848–1935), German Karl Erich Correns (1864–1933), and Austrian Erich von Seysenegg (1871–1962), had independently discovered the same laws of heredity. Searching the literature as they prepared to publish their work, they found, in 1900, that Mendel had preceded them. They gave him due credit as discoverer, but de Vries also described mutations, genetic alterations that can result in new traits.

CHROMOSOMES AND DNA

Interest in the puzzle of inheritance now shifted to the chromosomes. In 1902, American geneticist Walter S. Sutton (1877–1916) suggested that chromosomes might contain the genetic factors predicted by Mendel. During cell division, chromosomes behaved like Mendel's inherited traits. In 1909, American

Left: Fruit fly. Top left: Models of DNA. Top right: American geneticist Thomas Hunt Morgan. Research into how genes work included Morgan's experiments with fruit flies (Drosophila melanogaster) *and the discovery of the double-helix structure of DNA.*

geneticist Thomas Hunt Morgan (1866–1945), who used fruit flies for breeding experiments, discovered that genes are located on chromosomes in a fixed linear order. Morgan also explained sex-linked inheritance: some traits pass to only one or the other sex depending on whether the offspring receives two X chromosomes (female) or one X and one Y chromosome (male).

It remained unclear how genes determined the traits. Clues began to emerge in the 1930s and 1940s. In 1931, American geneticist Barbara McClintock (1902–92) showed that chromosomes can break and exchange parts during the formation of eggs or sperm. In the 1940s, American scientists

George W. Beadle (1903–89) and Edward L. Tatum (1909–75) discovered that a specific gene determines the production of only one enzyme. In 1944, American bacteriologist Oswald T. Avery (1877–1947) found that DNA alone—a component of chromosomes—determines heredity. But how it did so was unknown.

In 1953, American biologist James Watson (b. 1928) and British molecular biologist Francis Crick (1916–2004) discovered the double-helix structure of the DNA molecule, which showed how DNA transmits inherited traits from one generation to the next. Their work was indebted to X-ray studies of DNA done by British chemist and molecular biologist Rosalind Franklin (1920–58). The discovery of the structure of DNA unlocked the genetic code that governs heredity and made possible many new areas for research and innovation.

BIOTECHNOLOGY AND THE HUMAN GENOME

One path that Watson and Crick opened was genetic engineering, the manipulation of the genetic structure of organisms. In 1973, American biochemists Stanley Cohen (b. 1922) and Herbert Boyer (b. 1936) showed that this could be done through artificial recombination of genes, the pasting of genes from one organism into another. The technique since has been used to develop new medicines, crops, materials, and experimental organisms.

In the early twenty-first century, the Human Genome Project mapped the sequence of the entire human genetic inheritance. Impressive as this milestone was, it stood on the shoulders of many predecessors.

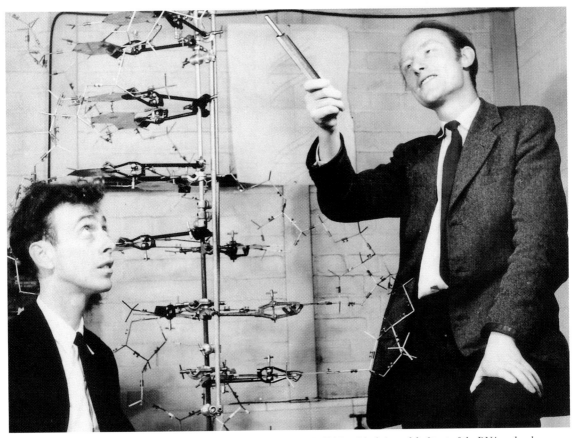

American biologist James Watson and British molecular biologist Francis Crick, with their model of part of the DNA molecule. As co-discoverers of the double-helix structure of DNA, they were indebted in their research to X-ray studies of DNA done by British chemist and molecular biologist Rosalind Franklin.

Explosion of Knowledge

The twentieth and twenty-first centuries have brought a tremendous increase in biological knowledge. As a result, biology has branched into many subdisciplines, each with its own specialized expertise and record of achievement—from immunology, the study of the immune system, to neuroscience, the study of how the nervous system functions. Yet all have connections to other disciplines and are informed by some common themes, particularly the Darwinian theory of evolution.

Paleontology is an example. Once, fossil hunters might have been happy simply with any find from any period. But now, informed by evolution and ecology, they specialize in particular species, lineages, or eras and try to understand the behavior and ecology of the animals involved. Many discoveries have flowed from this trend toward specializa-

tion. Research into human ancestry was sped by the mid-twentieth-century discovery in Africa of australopithecines and other hominids by British anthropologists Louis Leakey (1903–72) and Mary Leakey (1913–96). In 1968, American paleontologist Robert Bakker (b. 1945) transformed dinosaur research with his proposal that dinosaurs were warm-blooded.

ANIMAL AND HUMAN BEHAVIOR

Living species, too, began to receive more specialized kinds of attention in the twentieth century. Psychologists experimented with animal behavior in the laboratory, following the lead of Russian physiologist Ivan P. Pavlov (1849–1936). Pavlov showed that, through repeated association, a conditioned reflex could be created: an artificial stimulus (such as

a bell), if substituted for a natural one (such as food), could cause the same physiological reaction as the natural one (dogs trained to salivate at the sound of the bell). But other scientists, ethologists, studied animal behavior in natural environments, tracking them in the wild, observing their interactions, and trying to understand how their behavior could have evolved through natural selection. Austrian-born German zoologist Konrad Lorenz (1903–89) is considered the founder of ethology. Through studies of birds, he described the process of imprinting, by which a young animal may learn to identify its parent.

Some biologists began to consider whether human behavior might have evolutionary aspects. American entomologist Edward O. Wilson (b. 1929) founded sociobiology, which applies evolutionary principles to the study of animal social behavior, including humans. When proposed in the 1970s, his work and ideas were controversial, but since then an evolutionary approach to understanding human behavior has gained increasing acceptance.

ASTOUNDING DISCOVERIES

Modern biology has not only become more specialized; it has become more far-flung. The discoveries of the twentieth

Above: Russian physiologist Ivan Pavlov. Top left: A dinosaur's skull. Twentieth-century biologists made wide-ranging discoveries in a variety of subfields.

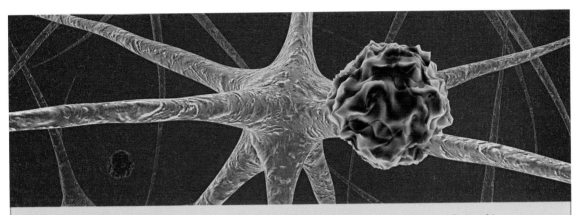

First identified in the late nineteenth century, viruses, although not alive, can cause disease in living things.

A NEW KIND OF MICROBE

As biology developed, any discovery could create a new subdiscipline. A classic example is the discovery of viruses, which led to the establishment of virology.

In 1898, Dutch microbiologist Martinus Beijerinck (1851–1931) discovered what he called a filterable virus (Latin for "poison")—a disease-causing particle so small that it could pass through a filter designed to keep out bacteria. Over the next few decades, biologists discovered that viruses could cause disease in plants, animals, humans, and even other microbes (bacteria). But viruses had strange characteristics that required special handling. For example, they could not be cultured in a nonliving medium, but they flourished in a living one. In the 1930s, scientists learned to grow them inside chicken embryos. As the field become ever more complex, virology was born. Its achievements included the development of the polio vaccine in 1955 by American microbiologist Jonas Salk (1914–95).

and twenty-first centuries have ranged from the microscopically tiny to the geographically remote. Early in the century, physiologists identified the first neurotransmitter, or chemical messenger of the nervous system, acetylcholine. Late in the century, marine biologists discovered previously unknown ecosystems of invertebrates living in the deep sea around hydrothermal vents, powered by underground heat and the symbiotic relationship with bacteria uniquely specialized to live in the environment.

Wherever they took place, many of the developments of the twentieth and twenty-first centuries have been astounding. In 1952, American chemists Stanley Miller (b. 1930) and Harold C. Urey (1893–1981) shed light on how life first originated when they produced amino acids, key components of life, from simpler chemical compounds. In England in 1978, the world's first successful human pregnancy by in vitro (test tube) fertilization came to term. In 1996, British scientist Ian Wilmut (b. 1944) cloned the first mammal, a sheep named Dolly, from an adult cell of another animal. As biological knowledge continues to explode, more such firsts can be expected.

British scientist Ian Wilmut with Dolly, the first mammal cloned from an adult cell.

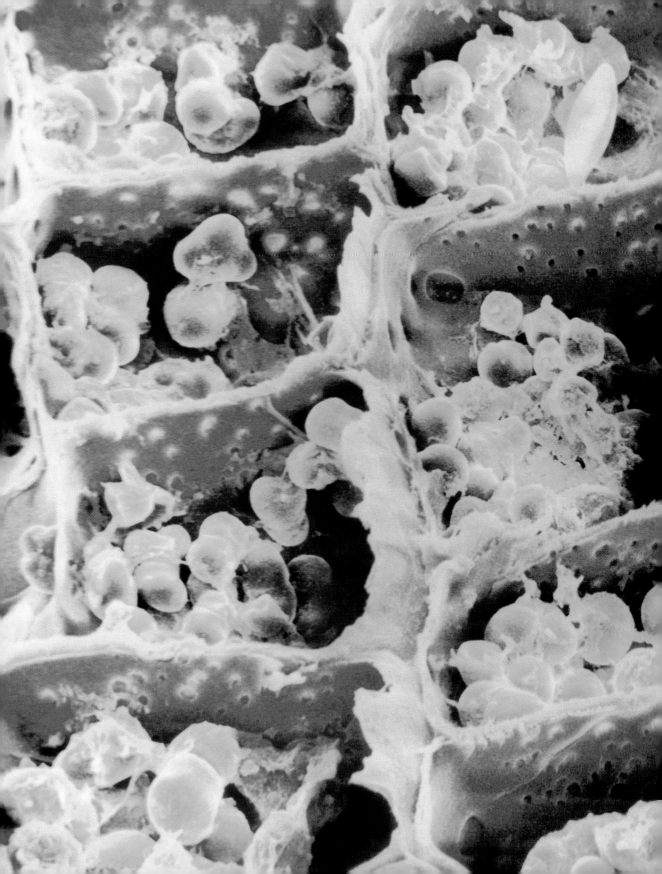

CHAPTER 4

THE BUILDING BLOCKS OF LIFE

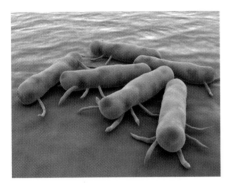

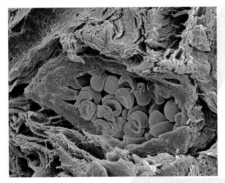

Left: Mimosa cells. Top: Salmonella bacteria. Bottom: Coronary artery and red blood cells. In every organism, cells are the building blocks of life. An organism may consist of one cell, such as Salmonella bacteria. Or it may consist of many cells, such as the tropical plants called mimosas. Cells are often specialized for specific functions, such as the oxygen-carrying red blood cells.

Cells are easy to overlook. Until the invention of the light microscope they went unnoticed because they are almost always too small to be seen with the naked eye. The period at the end of this sentence is big enough to contain about 500 typical cells, each measuring about 1/1,000 inch (0.0025 centimeter) in diameter. Yet despite their small size, cells are vitally important. They are the basic unit of all living things.

No organism—plant, animal, or otherwise—can be understood without grasping how it works at the cellular level. Cells are themselves alive, the smallest structures that can be described that way. Although built from inanimate molecules, each carries out the essential activities of life such as reproduction, growth, and metabolism. In organisms such as bacteria, these activities are carried out in a single cell. But other cells are also able to be organized into larger, many-celled organisms, from seaweed to blue whales, contributing to the life of the whole even while they carry out their own biological activities. Each human being is composed of more than 10 trillion cells working in this manner. This chapter will consider these eukaryotic cells.

What Makes a Cell?

Although cells are small, they are not the smallest items in nature. The latter include sub-atomic particles, bits of matter such as protons, neutrons, and electrons that combine to form atoms, which connect to make molecules, which ultimately combine to form cells.

An atom is so small that a human hair is more than a million times thicker than it is. The typical atom consists of a nucleus of protons and neutrons surrounded by a cloud of electrons. The atom remains stable through the attraction of the positively charged protons and the negatively charged electrons. Neutrons have no electric charge.

Atoms are the building blocks of chemical elements: simple substances such as carbon, hydrogen, nitrogen, and oxygen. Each element contains only one kind of atom and that atom is distinguished by its number of protons, which is equal to the number of electrons. Carbon, for example, has six protons, hydrogen one, nitrogen seven, and oxygen eight.

MOLECULES

Atoms unite with one another to form molecules, which are the building blocks of more complex substances called compounds. Examples include carbon dioxide (a compound of one carbon and two oxygen atoms) and water (a compound of one oxygen and two hydrogen atoms). What unites the atoms in a molecule are chemical bonds, linkages formed when two or more atoms share or transfer electrons. The carbon atom is unusual because it can form bonds in many different ways, giving rise to a tremendous number of compounds. The carbon atoms in these compounds bond with other carbon atoms and with atoms of other elements, forming chains and rings. Carbon compounds—consisting mainly of carbon with hydrogen, nitrogen, and oxygen in various proportions—are the core material of all cells and all living things.

Carbon molecules can be quite large and complex; they are then called macromolecules. Proteins, for example, are macromolecules formed of smaller molecules called amino acids, which themselves are composed of several distinct groups of atoms. Most simple proteins contain 100 to 300 amino acid units. Proteins are the principal building blocks of cells, making up 80 percent of the dry weight of muscle; some, called enzymes, also speed up chemical reactions within an organism.

MORE MACROMOLECULES

Other important macromolecules in the body are carbohydrates; these include sugars, which

Above: A human hair seen through an electron microscope. Top left: Artist's impression of electrons orbiting an atomic nucleus. Atoms, which each consist of a nucleus surrounded by orbiting electrons, form molecules, which form cells, which form structures such as hairs.

provide energy for cellular processes; starches and glycogen, which store energy; and cellulose, the main supporting material in plants. Nucleic acids, such as deoxyribonucleic acid (DNA), are macromolecules that carry instructions within the cell. Lipids such as fats and oils are macromolecules that can store fuel, serve as building material for cell membranes, and (if they are members of the class called steroids) serve as hormones that mediate the transmission of chemical signals.

Cells are built from macromolecules, which are built from smaller molecules, which in turn have been built from atoms. Yet, a cell never just arises from atoms. Cells are so complex that they can only arise from other cells through the process of reproduction, in which one cell divides to become two. Each cell contains organelles; structures that have specific tasks and are themselves made up of molecules. Nevertheless, cells are fundamentally made of atoms, a characteristic they share in common with all nonliving things.

Right: The organic, or carbon-containing, structure of a single caffeine molecule. A compound, caffeine consists of atoms of carbon (represented by the symbol C) in combination with hydrogen (H), nitrogen (N), and oxygen (O). In the symbol CH₃, the "3" represents three atoms of hydrogen bonded with one atom of carbon. Below left: Graphite. Along with diamonds and amorphous carbons, graphite is one of the three main natural forms of pure carbon.

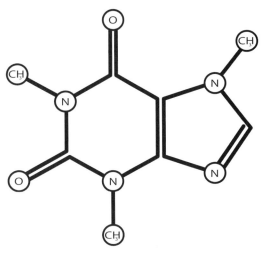

THE USES OF ORGANIC COMPOUNDS

Compounds that contain carbon in combination with one or more elements are traditionally called organic compounds because they were once thought to occur only in organisms. Although it is now known that they can be produced outside of them, many organic compounds still do come from organisms and are vitally important to other organisms, including humans.

Oil refinery. The hydrocarbons in petroleum fuel modern civilization, but must be processed before they can be used. Oil refineries convert crude petroleum into products such as gasoline.

An example is hydrocarbons, compounds containing only hydrogen and carbon. These include natural gas and petroleum, which derive from the fossilized remains of prehistoric organisms. Hydrocarbons are the fuel of modern civilization, from factories to cars, and the source of such everyday materials as plastics.

Most food products consist mainly of organic compounds such as carbohydrates, fats, oils, and proteins. These nutrients, or nourishing substances, are taken from plants, animals, and even fungi such as mushrooms. Organic compounds are also used to produce soaps, detergents, pesticides, and medicines.

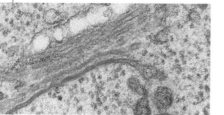

Inside a Cell

Cells come in many shapes and sizes, but all have certain similarities. Every cell is bounded by a plasma membrane consisting mostly of protein and lipid. Within the membrane is a jellylike material called cytoplasm. It is about 90 percent water, but also contains other substances such as proteins, sugars, potassium, and sodium, which make the cytoplasm viscous instead of watery. Within the cytoplasm are organelles,

structures with specific functions such as respiration or digestion. The entire cell operates as a kind of factory, taking in energy and using it to manufacture chemical products, either for its own use or for the use of the organism of which it is a part.

Eukaryotic cells—those of plants, animal, fungi, and protists such as amoebas—have a control center called a nucleus that is embedded in the cytoplasm yet set apart by its own membrane.

Unlike eukaryotic cells, prokaryotic cells (see chapter nine) do not have a nucleus.

THE NUCLEUS

The nucleus is the largest organelle in a cell. It is usually spherical and located near the center of the cell. It is bounded by the nuclear envelope, which is a double membrane that has pores to allow exchange of materials with the surrounding cytoplasm.

EUKARYOTIC CELL

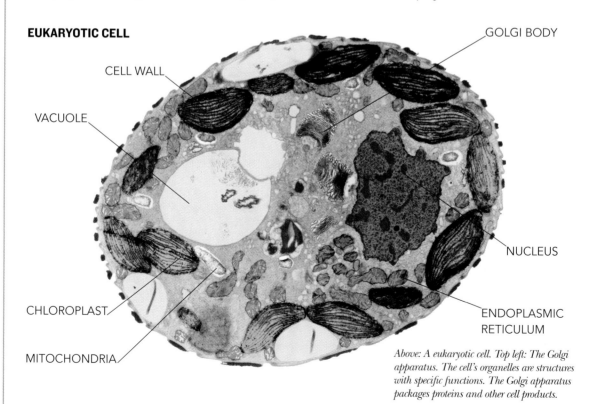

CELL WALL
VACUOLE
CHLOROPLAST
MITOCHONDRIA
GOLGI BODY
NUCLEUS
ENDOPLASMIC RETICULUM

Above: A eukaryotic cell. Top left: The Golgi apparatus. The cell's organelles are structures with specific functions. The Golgi apparatus packages proteins and other cell products.

A plant's leaves are green because its cells have chloroplasts, organelles that contain the green pigment chlorophyll.

Within the nucleus are the chromosomes, long, threadlike structures that contain DNA, the nucleic acid that makes up an organism's genes. The genes carry the information for cellular structure and activities, and also carry instructions for the products required for replication. The nucleus also contains round bodies called nucleoli, which form in some chromosome regions to facilitate the formation of the sites of protein production (ribosomes). Nucleoli are made of proteins and ribonucleic acid (RNA).

OTHER ORGANELLES

Other structures found within cells include:

- A mitochondrion, which is the largest organelle next to the nucleus. It is the powerhouse of the cell. There are hundreds or even thousands of mitochondria in a cell, each converting the chemical energy in food into a form of energy that the cell can use. A mitochondrion is surrounded by a double membrane and contains many folds covered with enzymes.
- Endoplasmic reticulum is a complex network consisting of folds of membrane that form flattened channels and tubular canals. It helps move materials through the cytoplasm and to the plasma membrane.
- A Golgi apparatus or Golgi complex is a stack of flattened membrane sacs that package proteins, lipids, and other products of the cell.
- A ribosome is a small, circular organelle; ribosomes are more numerous than any other organelles in the cell. Some are attached to the endoplasmic reticulum. They synthesize proteins that are used in the cell or that the cell secretes for use elsewhere in the body such as hormones and digestive enzymes.
- A lysosome is a round body containing enzymes that can break down food or foreign particles. The lysosomes inside white blood cells can destroy bacteria.
- A chloroplast is a lens-shaped organelle found only in plant and algae cells. Bounded by a double membrane, a chloroplast contains chlorophyll and carries out photosynthesis, the conversion of sunlight into food.
- A cytoskeleton is a network of protein rods that serves as the cell's skeleton, giving it rigidity and support. A cytoskeleton consists of hollow microtubules, or tubes, and solid microfilaments, or threads.

Computer model of the DNA molecule. Chromosomes contain these DNA molecules.

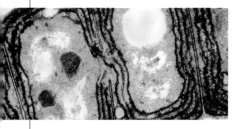

Guarding the Gates

A cell without a membrane is not a cell, just as a house without walls or roof is not a house. The plasma membrane separates the cell from its environment and is vital to some of its functions, such as making a physically separate copy of itself through reproduction. But a membrane cannot completely wall off the cell from its environment. Cells in an organism interact with other cells; they communicate by releasing molecules that the blood conveys to other cells and that serve as signals for action. Cell membranes also receive nutrients and release wastes. They serve as gates as well as walls—gates that are carefully guarded.

Many organisms have an actual cell wall. This is a rigid outer layer that surrounds the soft plasma membrane and gives protection and shape. The cells of animals do not have cell walls, but those of bacteria, algae, fungi, and plants do. In herbaceous plants—those with little permanent woody tissue such as roses and garden peas—cell walls give mechanical support.

HOW A CELL MEMBRANE WORKS

The cell membrane is a gate that guards itself, like a living gatekeeper. It is partially permeable or semipermeable (some things can pass across the membrane, while other molecules cannot), containing microscopic holes that allow the small molecules of water and of certain substances dissolved in water to pass through, but do not allow the passage of larger molecules or molecules of certain chemical composition.

Typically, materials are transported in and out of a cell by the processes of either passive transport or active transport. In passive transport, the cell does not expend energy to move molecules. Instead, molecules diffuse; they move unaided from an area of greater concentration to one of lesser concentration. For example, diffusion occurs when perfume molecules escape from a newly opened bottle and scatter the scent throughout the air. In

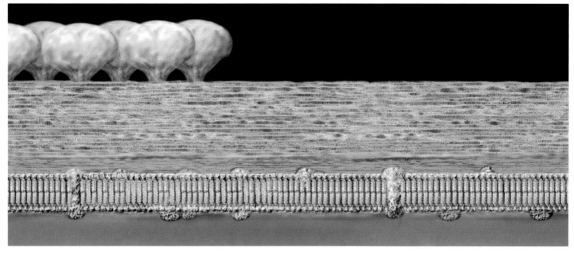

Above: Diagram of an archaebacteria cell membrane. Single-celled organisms, archaebacteria may have been among the first life forms.
Top left: Cyanobacteria, or blue-green algae, with cell walls. Cell membranes and cell walls permit selective interaction with the environment.

CELLULOSE

People do not eat grass. Grass cell walls are largely made of cellulose, a substance human beings are unable to digest. Cows, sheep, goats, and other ruminants can digest it because of their specialized digestive systems. Nonetheless, cellulose has a variety of uses for humans as well as for cows.

Cellulose is a polysaccharide, a macromolecule composed of many units of sugar. The plant cell synthesizes and secretes it through the cell membrane so it can form the cell wall, the rigid boundary of the cell. Plants, many algae, and some fungi have cell walls made primarily of cellulose. Although humans cannot digest cellulose, it constitutes the dietary fiber in foods such as vegetables, fruits, and nuts. As the digestive system eliminates fiber, it helps in the system's healthy functioning. Cellulose is also used to produce an array of materials, including paper, cotton, cellophane, and rayon.

Paper towels, like other paper products, are made from cellulose fibers tangled together and joined by chemical bonds.

Hydroponic plant. The roots of a plant, whether grown in nutrient solution or in soil, obtain water through osmosis.

the body, if the fluid outside the cells contains a higher concentration of oxygen molecules than the cells do, the oxygen molecules will tend to be transported by diffusion into the cells. Movement of water molecules into the body via diffusion is called osmosis. Some molecules are so large that they have a chemistry that restricts their diffusion across the cell membrane. They need help to diffuse into the cell.

In facilitated diffusion, protein carrier molecules help such particles to cross the cell membrane. For example, a glucose carrier protein, located inside the membrane, combines with a glucose molecule to shepherd it inside the cell.

ACTIVE TRANSPORT

In the active transport process, molecules move across a membrane or other barrier from an area of lesser concentration to one of greater concentration. Since this movement against a concentration gradient cannot occur spontaneously, an expenditure of energy is necessary.

This process is akin to the uphill movement of water; a pump is needed for the process. For example, to "pump" sodium ions (atoms with an electric charge) out of a nerve cell, a protein carrier molecule in the cell membrane combines with the sodium ions and uses energy to release the ions outside the cell.

Another form of active transport is phagocytosis, the process by which a white blood cell (also referred to as a leukocyte or immune cell), one of the body's defense cells, expends energy to engulf an invading bacterium. The white blood cell closes around the germ and destroys it with enzymes.

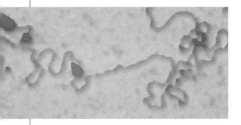

The Workings of Cells

Cells are factories that are constantly at work bringing in raw materials, using energy, making and sending out products, and releasing wastes. Yet, their varied activities have several things in common. They take place through the three-dimensional interaction of the surfaces of molecules. They require enzymes, energy, and water. Finally, they are part of an orderly pattern of metabolism that is essential to life.

MOLECULES
FITTING MOLECULES

Molecules are three-dimensional objects. Protein molecules, for example, can form flat sheets, globular masses, and even spiral chains. The structure of a molecule is what enables it to have and to perform a function. Enzymes are molecules that speed up chemical reactions in living things without being changed themselves. In chemical terms, this makes them catalysts. Made of protein, each enzyme precisely fits one substrate, the molecule upon which it works. For example, the stomach enzyme called pepsin breaks down proteins in food into peptides. Pepsin can do this because its molecular shape fits the food-protein substrate and weakens its chemical bonds.

Even the shapes and sizes of cells are related to their functions. The cells of skeletal muscles—the muscles that control a vertebrate's body movement—are shaped like long, thin cylinders called muscle fibers. Their structure is what enables them to contract and cause body motion.

REQUIREMENTS

For its various activities, a cell has certain requirements. Enzymes are needed to speed chemical reactions. Energy is also needed to fuel reactions, in particular, the molecule adenosine triphosphate (ATP). Just as modern-day household appliances require electric current and will not run directly on burning wood, the body's cells require the energy stored in food to be converted first into molecules of ATP, which they can break down to release energy for their operations.

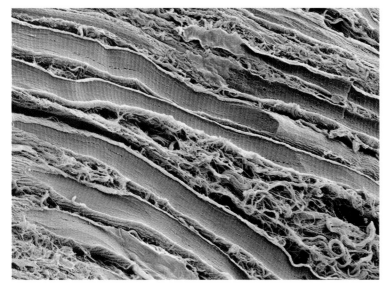

Left: Skeletal muscle fibers. Above: A radioisotope scan of thyroid gland. Top left: Plasmid and enzymes. Cells are constantly at work performing essential functions. Skeletal muscles are built to contract, causing motion. The thyroid gland controls the rate of metabolic processes. Plasmids and restriction enzymes are structures in bacteria.

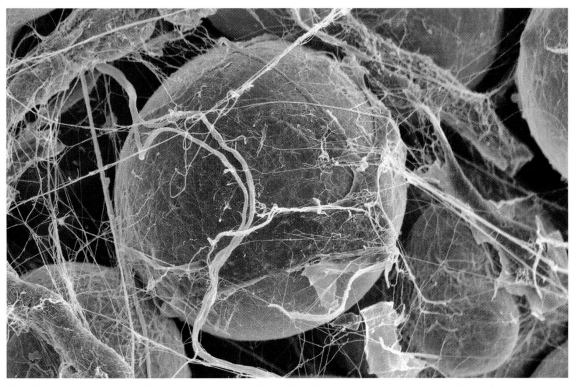

Fat cell. One of the body's anabolic reactions is to combine fatty acids with glycerol to create fat, a reserve of energy for lean times and an insulation against heat loss. The fat is stored in fat cells, among the human body's largest cells.

Water is also vital. Making up about 70 percent of the human body's weight, water is required for most of the chemical reactions that happen in the body. It transports nutrients, enters into many reactions, and dissolves many of the substances that enter into reactions.

METABOLISM

The activities of cells are part of metabolism, the sum of the chemical reactions by which cells perform their activities.

Two phases of metabolism are constantly performed: catabolism and anabolism. Catabolism is the breaking down of molecules with the release of energy. Anabolism is the building up of complex molecules from simpler ones with the required use of energy. In both catabolism and anabolism, the chemical reactions typically follow a metabolic pathway, a sequence of small steps.

In the catabolic reaction known as cellular respiration, cells break down the simple sugar glucose (obtained from food) to produce ATP, the form of energy that cells can use. Respiration occurs in two stages. In the first stage, glycolysis, which does not require oxygen, glucose is broken down into pyruvate. In the second stage (the Krebs cycle) pyruvate is broken down to produce ATP in an oxygen-requiring reaction, with carbon dioxide and water generated as waste products. Energy is stored in the chemical bonds that link the atoms of ATP.

In anabolic reactions, cells break the chemical bonds in ATP and release the stored energy necessary to assemble larger molecules. Amino acids are joined to form proteins to repair and extend tissues and serve as enzymes or hormones. Molecules of glucose are combined to form the storage compound glycogen, and fatty acids are combined with glycerol to form body fat. More food comes into the body, and the cells break it down for use in more molecular assembly, ensuring that life continues.

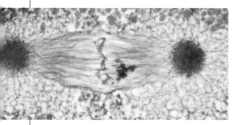

The Magic of Self-Copying

Reproduction is one of the characteristics that distinguish cells as living things. A cell reproduces by dividing—splitting to form two daughter cells.

Reproduction is the beginning and end point of the cell cycle, the life trajectory that cells undergo just as animals pass through their own life cycles from birth to death. Each cell begins life as the daughter of a parent cell, subsequently spending most of its life growing and working: a period of time called the interphase. During interphase the DNA molecules in the nucleus replicate, or self-copy, so that the nucleus ends up containing two copies of each

Below, from left: The stages of mitosis, a common form of nuclear division: prophase, metaphase, anaphase, telophase. Top left: Metaphase.

chromosome rather than the one copy with which it started. When that process is finished and the cell is large enough to divide, it splits into daughter cells, each with one copy of the chromosomes, and the cycle begins anew. As a result of the cell cycle, bodies grow and worn or damaged tissues are replaced. The duration of the cell cycle varies in different tissues. As an example, the epithelial cells of the intestine wall may divide every eight to ten hours.

CELL DIVISION

Cell division occurs in two phases. The first is nuclear division, the time when a nucleus splits. In the second, cytokinesis, the cytoplasm divides. When cytokinesis is complete, the cell has split into two cells, each with its own nucleus.

In eukaryotic organisms, nuclear division is essential to reproduction because the chromosomes in the nucleus contain the hereditary material (DNA), which encodes cellular function, growth, and division. Human red blood cells do not have nuclei and so die (after a lifespan of about 120 days) without undergoing cell division. Bone marrow replenishes the body's supply of red blood cells by constantly manufacturing them.

NUCLEAR DIVISION

Nuclear division occurs in two ways. The first and more common way is mitosis, in which the nucleus divides and forms two identical nuclei. Mitosis occurs in four phases:

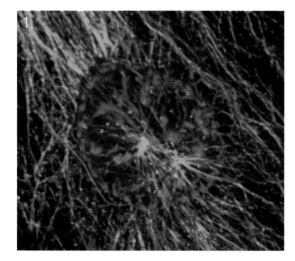

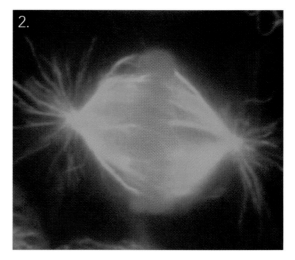

CANCER: MITOSIS RUN AMOK

Mitosis normally occurs in an orderly way. Skin cells divide often enough to replace dead skin, but no more. Bone cells divide often when a child is growing, but the bones do not keep growing afterward. When cancer occurs, however, cells behave wildly. They multiply without restraint: mitosis run amok. Their runaway growth crowds normal organs, interfering with their functioning. Furthermore, the cancerous tissue may spread throughout the body, causing death.

Altered (mutated) genes are the cause of cancer. A mutated gene can convey the wrong instruction, causing unintended cell growth either by a direct signal or indirectly, by suppressing another mechanism that normally stops cell growth. The mutation can be caused by errors in DNA replication over time. It can also be inherited or caused by environmental factors such as cigarette smoke and ultraviolet light.

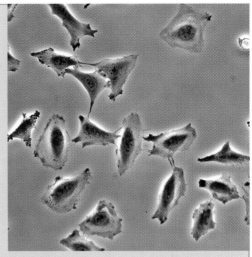

Cells derived from a cancerous cervix. Genetic mutations give cancer cells an unusual, disordered appearance.

1. **prophase**—the chromosomes divide along their length; each becomes a pair of identical strands known as chromatids, held together by a centromere
2. **metaphase**—the membrane around the nucleus breaks down
3. **anaphase**—the centromeres split and the paired chromatids move apart
4. **telophase**—the chromatids collect at opposite sides of the cell; a new nuclear membrane forms around each group, producing two daughter nuclei

After nuclear division is complete, cytokinesis occurs. In animals, the cytoplasm in the middle of a cell turns inward, pinching the cell into two new cells. In plant cells, a cell plate of cellulose forms between the two new cells.

The second form of nuclear division is meiosis, in which the cell divides into four daughter cells, each of which have only half the number of chromosomes in the parent cell. The result is the production of reproductive or sex cells, also called gametes. Each of these is capable of uniting with another gamete to generate a fertilized egg cell, which can grow into a whole new organism.

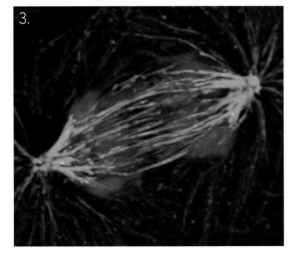

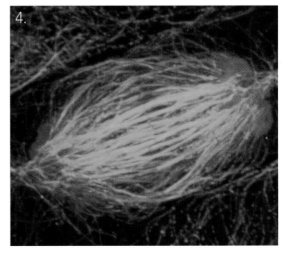

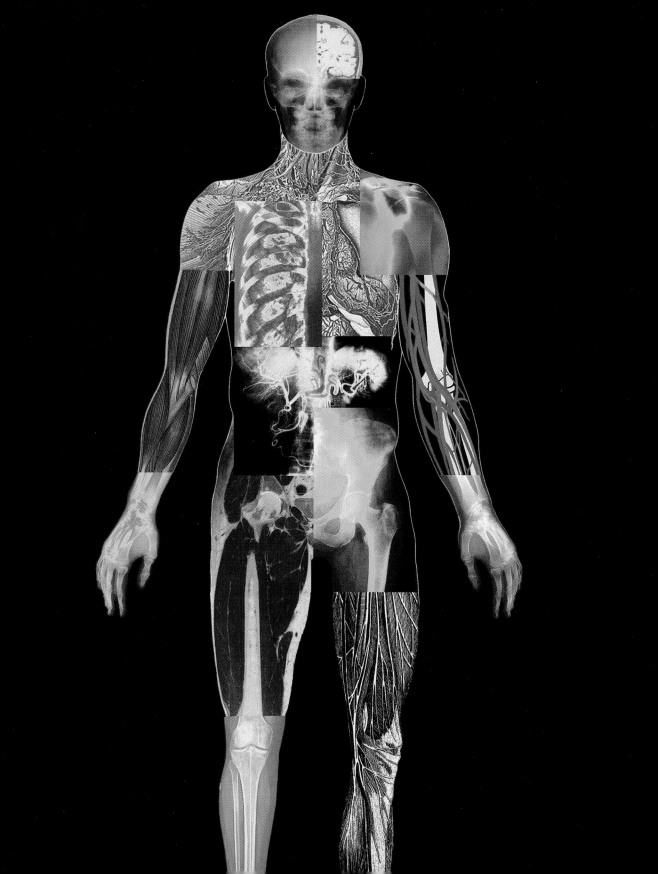

CHAPTER 5

A COMPLEX ARCHITECTURE

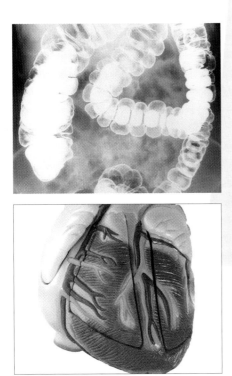

Left: A composite image of the human body. Above: X-ray image of the small and large intestines. Bottom: Model of the human heart, showing ventricles and major vessels. Anatomy, the study of the structure of living things, employs a number of methods in order to understand how organisms are constructed. In the composite image (left), magnetic resonance imaging (MRI), X-rays, and dissection are used.

In many living things, each cell forms part of a larger and more complex whole. Like bricks in a mansion, every cell in a multicellular (many-celled) organism contributes to an impressive architecture with an array of parts. Anatomy, the branch of biology concerned with the structure of living things, examines the complex architecture of organisms. This study is a sister to physiology, the branch of biology that investigates how living things use that architecture to function (see chapter six).

Certain anatomical features are characteristic of most multicellular organisms. Cells are typically organized into specialized tissues and tissues into organs. In turn, organs often are grouped into organ systems that work together to perform a still larger function, such as circulation. Other anatomical structures are specific to particular types of organisms. Leaves and flowers are common organs in plants; hearts and eyes in animals. Plants, which are nourished by sunlight, have no need of the digestive systems that are characteristic of animals. Vertebrate animals, which are able to move freely through means such as their musculoskeletal systems, are unhindered by the roots that characterize plants. Anatomy is as varied as the ways in which organisms live.

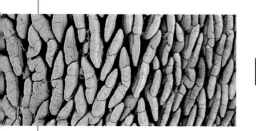

From Cells to Bodies

In multicellular organisms, cells are generally organized into three levels of increasing order: tissues, organs, and systems.

TISSUES

A tissue is a collection of cells. These cells, along with the intercellular material between them, are similar in structure and work together to perform a particular function. For example, nervous tissue in animals is specialized to conduct nerve impulses throughout the body.

In animals, there are four primary types of tissue:

- Epithelial tissue consists of closely packed cells in a sheet with little substance between the cells. The functions of epithelial tissues are to protect and secrete, or produce specific substances. Epithelial tissue covers the outer layer of skin and lines internal body surfaces such as those of the digestive tract and blood vessels. It also forms glands and parts of sense organs.

- Muscle tissue consists of sheets or bundles of cells that can contract, producing movement. It comes in three forms: voluntary muscle, which produces chosen movement; involuntary muscle, which produces automatic movement, such as that of the intestine; and cardiac muscle, found only in the heart.

- Connective tissue, made of cells and fibers, lies within a large amount of intercellular material known as a matrix. The functions of connective tissue include support, packing, and defense. Connective tissues include bone, cartilage, blood, and lymph.

- Nervous tissue consists of neurons, or nerve cells, which

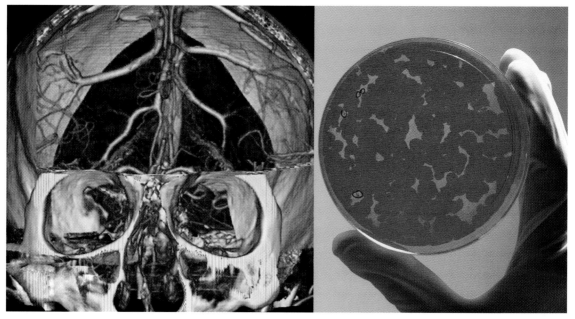

Above left: MRI of human brain. Above right: Cultured epithelial cells. Top left: Intestinal lining. Cells, such as epithelial cells, are organized into tissues, such as the intestinal lining, which form organs, such as the brain.

conduct electrochemical impulses, and supporting cells called glia, which perform such activities as providing nutrients for neurons. The function of nervous tissue is communication, control, perception, and response to stimuli.

Plants have two major kinds of tissue: meristematic and permanent. Meristematic tissue, or meristem, consists of cells that divide rapidly and continuously, giving rise to cells that will differentiate to perform various functions. The growing tips of shoots and roots contain meristems. Permanent tissues are cells that have matured to take on special functions. These include epidermis (the plant's protective covering), parenchyma (which stores food and water), xylem (which conducts water and minerals upward from the roots), and phloem (which conveys food throughout the plant).

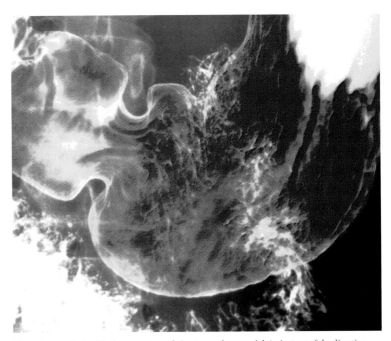

X-ray image of a healthy human stomach (center and upper right). As part of the digestive system, this curved, saclike organ stores and partially digests food before passing it on to the small intestine (upper left). Epithelial tissue, a protective lining, is found in the digestive tract.

Green sprout. The growing tips of a young plant's shoots and roots contain meristematic tissue or meristem, which consists of cells rapidly and continuously dividing.

ORGANS AND SYSTEMS

Just as cells form into tissues, tissues often form into organs. An organ joins two or more kinds of tissues to perform specific functions. For example, the human heart, which pumps blood throughout the body, contains muscle tissue, nerve tissue, and connective tissue. The lungs, stomach, kidneys, skin, wings, leaves, and roots are other examples of organs that consist of several types of tissue.

An organ system, as found in more complex animals, is a group of organs that carries out a major activity. The human digestive system, which breaks down food into a form that cells can use, includes the mouth, esophagus, stomach, small intestine, large intestine, gallbladder, liver, and pancreas. The human body also includes the respiratory, circulatory, nervous, urinary, reproductive, endocrine, skeletal, and muscular systems. In an organism that has organ systems, all the systems together compose the body.

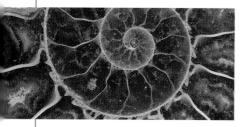

Skins and Skeletons

Just as a cell needs a membrane, a multicellular organism needs an outer layer to define its border. Whether this layer is of skin or of bark, it protects the organism from injury, infection, water loss, and other threats to its survival.

Many organisms also have other kinds of protective structures: hard parts, such as skeletons, shells, and teeth. In many cases, these hard parts afford not only protection, but also support for the internal organs, to keep them from collapsing on top of one another. In vertebrates, the skeleton also provides a rigid framework of levers that facilitates muscular action.

THE OUTER LAYER

The outer protective layer varies among organisms. Many plants have a thin outer layer of cells called an epidermis. In most trees and shrubs, the outer covering is bark, a substance that gets its toughness from a dead tissue called cork. Under their bark, trees and shrubs have wood: hard layers of tissue that get their stiffness from a chemical called lignin.

The outer layer that covers humans and many other mammals is skin. This consists of two layers. The outer layer is the epidermis and the inner, thicker layer is the dermis. The epidermis shields the organism from water because its outermost cells contain a tough, waterproof substance called keratin. In many animals, hair, feathers, hooves, horns, claws, and scales, all of which are rich in keratin, further protect the epidermis. The dermis contains blood vessels, nerves, and glands.

HARD PARTS

Some organisms are protected by hard external armor. Many marine animals including clams, oysters, and lobsters grow shells made of calcium carbonate. On land, snails and turtles also grow shells.

In many cases, the shells of animals are part of a rigid, external framework called the exoskeleton. The exoskeleton not only protects and supports the animal, but also provides points of attachment for muscles. This is the case for the arthropods, a category that includes crustaceans, insects, and arachnids. The exoskeleton of arthropods gets its toughness from a stiff, horny substance called chitin.

Left: Magnified skin cells. Right: Tree bark. Top left: Spiral shell. Organisms have many kinds of coverings to protect them from their environment. Skin, found in humans and other mammals, is a soft covering that consists of two layers, epidermis and dermis. The bark of trees and shrubs, formed from a substance called cork, is tough but not as hard as the wood underneath. The calcium carbonate shells of marine animals are even harder.

Spines of a sea urchin. Echinoderms such as the sea urchin have hard, spiny exteriors.

INNER SKELETONS

In many organisms, the hard parts are inside, taking the form of an endoskeleton, or inner skeleton. Indeed, vertebrates (including fish, amphibians, reptiles, birds, and mammals) are named for the central part of their skeleton, the vertebrae, a column also known as the spine or the backbone. The basic material of the skeleton of most vertebrates is bone.

Beetle. Like other insects, beetles have exoskeletons, rigid external frameworks made of chitin, a stiff substance.

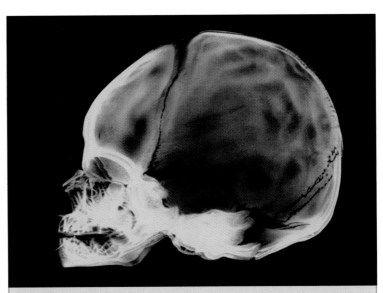

X-ray of baby's skull, showing anterior fontanel (dark area at upper center).

FINISHING THE SKULL

As any parent knows, the skull of a newborn is unique. An infant's head has soft spots—areas called fontanels. Though made of tough membrane, fontanels feel delicate compared to the hard skull around them.

Fontanels are the solution to an evolutionary problem: how to push a big-brained animal through its mother's narrow birth canal. If a baby were born with a full-size adult brain, its mother would need a gargantuan pelvis. But if the baby were born with a much smaller brain and a hard-boned skull, there would not be enough cranial room for later brain growth.

The solution: humans are born with brains that are only 23 percent of their eventual adult size, and with skulls that are not fully formed. The soft fontanels, which allow for rapid growth, generally close up by 18 months, but the skull bones do not fully knit together until adulthood.

It consists mainly of the minerals calcium, phosphate, and carbonate. The rest is organic material, mostly a fibrous protein called collagen. Embedded in bone are cells that help produce more bone. The spine forms the central axis of a vertebrate's body. At one end is the skull and at the other end the tail, or—in tailless animals such as humans—the coccyx. In vertebrates with limbs, the bones of legs, wings, or arms attach to the spine. Some vertebrates, such as sharks, have skeletons made not of bone but of cartilage, a waxy, rubbery substance.

In animals with skeletons, some hard parts may protrude outside. In echinoderms such as starfish and sea urchins, spines from the internal skeleton stick out through the skin. The cores of cow horns and deer antlers are extensions of the animals' skulls. And the hardest parts of all—teeth—are visible whenever the mouth is open.

Nerves and Muscles

To take in nourishment and avoid harm, organisms need to be able to respond to their environments. It helps if they can also move, whether to chase food or flee predators. Their anatomies present a variety of stratagems to carry out these functions, from the flagella of bacteria to the bending of plants toward sunlight. But the most complex approaches to irritability, or sensitivity, are found in the nervous and muscular systems of animals. Only animals—not plants, fungi, protists, or bacteria—have evolved a nervous system as a means for communication, control, and response to the environment. And only animals have evolved a muscular system as a means of motion.

RESPONDING WITHOUT NERVES OR MUSCLES

Organisms without nervous systems are still sensitive to their environments. Even unicellular, or single-celled, organisms can receive a stimulus and respond to it, but the reception and response happen within that single cell. The responses can be remarkable. In response to adverse conditions such as drying or freezing, the cytoplasm of some bacteria secretes a thick cell wall that surrounds the microbe while it enters a resting stage. The endospore (as it is known in

that stage) can survive for long periods, even perhaps thousands of years.

Plants and certain other organisms (such as sponges) are attached to one spot and are not capable of locomotion, or moving from one spot to another. But some organisms are capable of locomotion even without the muscular systems of animals. Some bacteria have flagella, thin hairlike structures that rotate from their base, propelling the microorganisms. Although plants are incapable of locomotion, they are capable of tropism, or growing toward or away from a stimulus. Plants bend toward sunlight when the light stimulates differences in the growth rate of its tissues.

NERVES

The nervous system of animals is a network of nerves that transmits information between sensory cells and organs, such as eyes and ears, and effectors, such as muscles and glands. Sensory organs contain receptors, cells specialized to detect a stimulus such as light or sound. Effectors are specialized to respond to stimuli.

In some animals, such as echinoderms, the nervous system is as simple as a nerve net in the body wall. But in most, the nervous system has two parts: the central nervous system, which coordinates and controls all the neural functions, and the peripheral nervous system, which

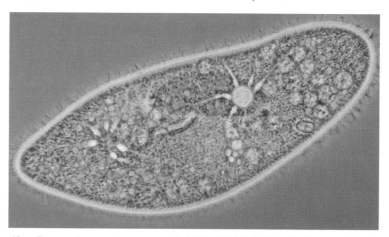

Above: Paramecium caudatum, *magnified 160x. Top left: Seedlings bending toward sunlight. Even without benefit of nerves or muscles, organisms can respond to their environment. Paramecia use cilia, hairlike projections covering their single cell, for locomotion and feeding. Plants bend toward sunlight when it stimulates differences in the growth rate of their tissues.*

connects the central nervous system to the rest of the body.

In invertebrates, or animals without a backbone, the central nervous system may be nothing more than a few nerve cords or bundles of nerve fibers. In vertebrates, the central nervous system consists of the brain (the main coordinating center) and the spinal cord (nervous tissue enclosed within the vertebral column). Animal brains vary in complexity. The human brain, with its billions of neurons, is the most complex of all.

The central nervous system of a cat is made up of the brain and the spinal cord, which extends to the tip of the tail. The peripheral nervous system serves the limbs and organs. This complex network transmits information between sensory organs and effectors, such as muscles and glands.

MUSCLES

Many animals have muscles, some more than others. Grasshoppers have about 900 muscles, compared to the fewer than 700 of humans. In invertebrates with exoskeletons, such as insects, the muscles are attached to the exoskeleton; in those without, such as squids, the muscles are unanchored. In vertebrates, many of the muscles are attached to the skeleton, forming a musculoskeletal system well suited for tasks such as walking, running, flying, lifting, and crushing.

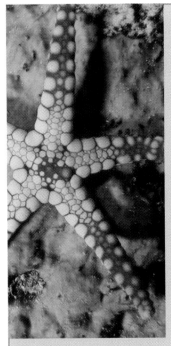

The starfish, or sea star, has no brain.

A STAR-SHAPED NERVOUS SYSTEM

Although animals in the phylum Echinodermata have a generally simple design, some of its varieties have intricately arranged nervous systems. For example, the starfish or sea star has no brain, but senses the world by mechanisms in its middle and in its arms (often five, arrayed in a star shape). At the middle of the starfish is a central disk with a ringlike nerve cord that connects additional nerve cords suspended in the grooves of its arms. The starfish senses light through a small, colored eyespot at the end of each arm. Other receptor cells also allow it to respond to smell and touch. They react to loss of limbs by regenerating replacement parts. Sometimes, an entire new starfish will form from a detached arm.

Channels for Air and Blood

On the surface, a behemoth like a rhinoceros or redwood tree seems immensely stronger than a bacterium. But the largest multicellular organism has some of the same weaknesses as the smallest unicellular one because it too is made up of cells. And each of those cells can only survive if it receives needed chemicals and disposes of wastes. The multicellular organism has an added problem: it must communicate efficiently with the other cells in the organism by sending and receiving messages. Multicellular organisms have evolved respiratory and circulatory systems that deal with these problems.

RESPIRATORY SYSTEMS

Respiration is the process by which living things get the oxygen they need and lose the carbon dioxide they do not need. Except for certain microorganisms, all living things require oxygen, just as fire needs oxygen to burn fuel. In living things, the fuel is pyruvate, an organic acid that the body produces from food. Through a sequence of chemical reactions that occurs within each cell and that requires oxygen, pyruvate is broken down. These reactions yield adenosine triphosphate (ATP), the molecule that provides the necessary energy for the cell's functioning.

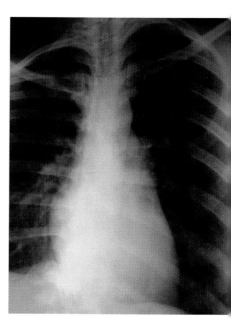

Above: Pair of rhinoceros. Right: X-ray of human lungs. Top left: Lime leaf, with stomata (white spheres). All multicellular organisms, from rhinos to plants to humans, are composed of cells that require support through respiration and circulation.

Green lizard. In vertebrates that breathe air, like this lizard, lungs are the organs that draw in oxygen and expel carbon dioxide. Their blood is pumped by the contractions of a heart.

The waste products of these reactions (known as cellular respiration) are carbon dioxide and water. To supply their cells with oxygen and get rid of the carbon dioxide (organismic respiration), many multicellular organisms have respiratory systems. These systems bring in oxygen from the outside and exchange it for carbon dioxide.

Oxygen enters through pores on the plant's surface—especially stomata, tiny openings in the leaves—and spreads throughout the tissues. It diffuses through the spaces between cells or is dissolved in tissue fluids. But in many animals, respiratory organs do the work of exchanging gases. The surface membranes of these organs are thin and moist, with an abundant blood supply. In vertebrates that breathe air, the organs are lungs, which draw in oxygen-rich air from the atmosphere and expel carbon dioxide. In fish and shellfish, the organs are gills, which take in dissolved oxygen from the surrounding water and release carbon dioxide.

Insects have a system of air tubes called tracheae, which transport air from the environment directly to every part of the body.

CIRCULATORY SYSTEMS

To transport oxygen to the body's cells and to take away carbon dioxide, many multicellular organisms have circulatory systems. Circulation is the flow of materials throughout the body, but circulatory systems transport much more than just oxygen and carbon dioxide. They also deliver nutrients to the body's tissues and carry away their wastes. They may also transport hormones, chemical substances that affect the functioning of cells and therefore aid in communication and control. Circulatory systems may also help regulate the body's temperature by absorbing and redistributing heat from cellular activities. And they may carry substances that assist in fighting injury and infection.

In vertebrates such as humans, all these materials are transported in a fluid tissue called blood.

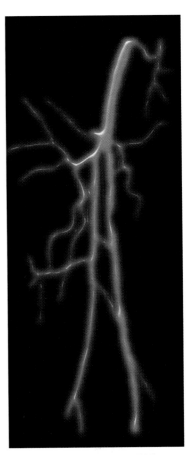

Femoral artery inside human thigh.

It is pumped by the contractions of the heart, a hollow, muscular organ, and conveyed by blood vessels, channels that branch to every tissue of the body. The lymphatic system, a network of small vessels that transports fluids from body tissues to the bloodstream, also is part of the circulatory system. The lymphatic system aids in fighting infection as well.

Many invertebrates, such as insects, also have blood, but in an open circulatory system that lacks blood vessels. Instead, the internal organs are suspended in blood, with materials exchanged in that active cavity.

A Place for Food

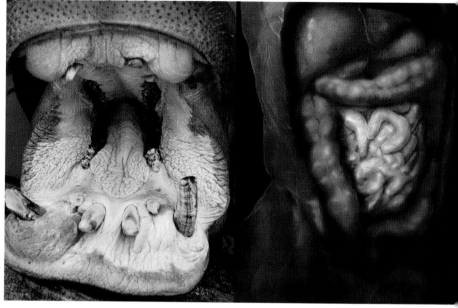

Plants have it easy. All they need is sunshine, air, and soil to make their food, through the process of photosynthesis. Other organisms, including all animals, have to ingest, or eat, food, which typically consists of the tissues of plants or animals. That food is usually made up of large molecules too complex for the eater's own cells to use directly. And it often contains many materials that are no use at all. To digest food—break it into simpler molecules and eliminate the indigestible materials—animals have digestive systems.

Above: Hippopotamus mouth. Right: Human small intestine. Top left: Flatworm, mouth in center. At its simplest, a digestive system is just a tube with at least one opening, a mouth, but it often has specialized organs such as the small intestine.

DIGESTING FOOD

At its most basic, a digestive system is just a tube that has at least one opening (a mouth) through which food can enter. That tube, known as a gut, digestive tract, or alimentary canal, contains digestive enzymes, molecules that speed up the chemical reactions that break down food.

Once the food is broken down into usable nutrients, cells lining the gut absorb them and pass them on to the rest of the body. Unusable materials exit through another opening, an anus. Some organisms do not even possess a second opening. For example, flatworms have only one opening, a mouth, through which they both ingest food and discharge solid waste. But in most animals, the alimentary canal extends from the mouth to the rectum, and one of its tasks is to move food along from the front to the back in an orderly way.

DIVISION OF LABOR

In vertebrates, the alimentary canal is divided into functional areas; each area performs one of the steps of the digestive process:

- The front part, the mouth, is responsible for receiving food. Specialized structures, such as teeth and saliva, begin the task of breaking down food before passing it to the next area.
- The next section, the esophagus, is a muscular tube that conducts food to the next region for processing.
- This region, the stomach, is a storage area where food is mixed and broken down by

hydrochloric acid and enzymes. Being able to store food before fully digesting it allows a large meal to be eaten at one time.

- The stomach sends the processed food into the small intestine, where chemicals and bacteria complete the digestion of food into component nutrients, which are then absorbed into the bloodstream.
- The last structure, the large intestine, is responsible for eliminating the indigestible parts of food. The large intestine absorbs water from the mass, so that it emerges from the anus as a solid waste, called feces.

BEYOND THE ALIMENTARY CANAL

In mammals such as humans, the digestive system includes not only the alimentary canal but also accessory structures such as the pancreas, liver, and gallbladder. These specialized organs evolved as outpouchings from the alimentary canal.

The pancreas releases pancreatic fluid, which contains digestive enzymes, into the small intestine. The liver discharges bile, a digestive fluid that aids in digestion of fats, into the small intestine. The gallbladder stores extra bile.

Some bodily wastes, however, are not eliminated through the digestive system. These wastes are not indigestible parts of food but are the products of the body's metabolic activity, such as excess water and salt and the nitrogen-bearing compound urea. The kidneys convert many of these substances to the fluid waste urine, which passes out of the body through a tube called the urethra.

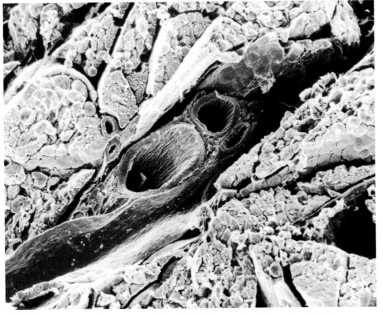

Electron micrograph of a pancreas. Part of the human digestive system, the pancreas pours juice containing digestive enzymes into the small intestine.

Chemical Messengers

Above: Monarch butterfly just emerged from its chrysalis. Top left: Canadian blueberries. Many phenomena of growth and change, from the branching of plants to insect metamorphosis, depend on hormones. The human hormone insulin closely resembles the same hormone in pigs.

The nervous system is one of two ways in which communication and control takes place in an animal. Another is the endocrine system, a system of tissues and organs that secrete chemical substances called hormones into the blood. These specialized body parts are glands that synthesize and secrete a specific chemical.

Some glands, such as sweat glands, secrete chemicals through ducts, or tubes; these are known as exocrine glands. But the glands of the endocrine system—mostly organs shaped somewhat like acorns—are ductless glands; they secrete their hormones directly into the bloodstream. Serving as chemical messengers, these hormones circulate through the blood to the other tissues or organs that are their targets. There, hormones modify the target's activity, usually in some fashion that is beneficial to the animal, such as making it grow, develop sexually, use digested food, or prepare for fight or flight.

PLANT HORMONES

Plants have hormones too, but they differ from animal hormones. They are not pumped out by glands or specialized hormone-producing organs; instead, the hormones are mainly synthesized in actively growing parts, such as the tips of roots and stems. And the hormones are not fitted to a particular target, but can affect virtually every tissue.

Plant hormones, which primarily regulate growth, include:
- Auxins, which cause plant stems to grow toward light
- Cytokinins, which control cell division and differentiation of plant tissue into roots, leaves, and other parts
- Gibberellins, which in many plants, stimulate growth, regulate blossoming, and spur the growth of seeds and buds after dormancy, or inactivity

THE VERTEBRATE ENDOCRINE SYSTEM

In the animal kingdom, invertebrates have hormones just as vertebrates do. For example, hormones control the development of insects such as bees and butterflies. But vertebrates, particularly mammals, have endocrine systems that are particularly similar to one another. They produce most of the same hormones, many of them nearly identical in structure and function.

Insulin, a pancreatic hormone that regulates the body's use of sugar, is structually and functionally so similar in pigs as it is in humans that doctors long used pig insulin to treat human patients with Type 1 diabetes, a disease in which the body does not produce enough insulin.

Since both the endocrine and nervous systems are involved in communication and control, it is not surprising that they interact, forming at their connecting points a neuroendocrine system. In humans, the main link is between a part of the brain called the hypothalamus (a small area at the brain's base) and the pituitary gland, a pea-sized gland underneath it. Through the narrow pituitary stalk that connects the two, the hypothalamus controls the pituitary with secretions called releasing hormones, which stimulate the pituitary to release its own hormones.

In humans, the most important endocrine glands are:

- The pituitary gland, which stimulates other endocrine glands, stimulates milk production and also regulates growth.
- Adrenal glands are two glands that regulate the use of digested foods, help the body adjust to stress, regulate excretion of sodium and potassium, and regulate sex characteristics.
- The thyroid gland regulates the rate of metabolic processes in the body.
- Parathyroid glands are four glands that control the amount of calcium in the blood.
- Sex glands, or gonads, are paired organs (testes in males, ovaries in females) that produce reproductive cells; they also release hormones that control secondary sexual characteristics.

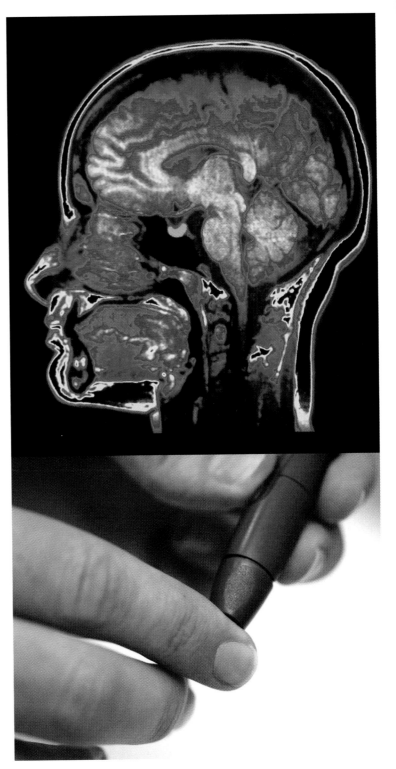

Right: Diabetic person takes a blood sample to monitor glucose levels. Above right: MRI scan of human brain, with pituitary gland highlighted in green. Human health depends on proper functioning of glands. These include the pituitary, which regulates growth, and the pancreas, which secretes insulin, regulator of sugar metabolism. Insulin deficiency is one cause of diabetes.

Where the Future Begins

All the bodily systems described so far have been dedicated to the survival of the individual organism. But if a species is to have a future after the individual's death, there must be a system that generates offspring. This system is the reproductive system.

WITH OR WITHOUT SEX

There are two principal forms of reproduction: asexual and sexual.

In asexual reproduction, all of the genes come from only one parent. In sexual reproduction, the genes come from two parents of opposite sexes. Unicellular organisms most commonly reproduce asexually, through the division of the organism's single cell. Some multicellular organisms—such as sponges, certain flatworms, and many plants—can also reproduce asexually, though all of these life forms may also reproduce sexually. Nonetheless for most animals, including humans, only sexual reproduction is possible.

THE BASICS OF SEX

The simplest approach to sexual reproduction is conjugation, in which two unicellular organisms form a cytoplasmic connection and exchange nuclear material. Through this means, bacteria, paramecia, and algae can have sex, even though they usually reproduce asexually.

In multicellular organisms, sex is more complicated. There is great variety in the architecture of sexual reproductive systems from mollusks to mammals, yet some basics similarities exist. In nearly all cases, individuals are differentiated into males and females, with each producing a distinct kind of gamete, a cell with only half of the chromosomes proper to that species. The male's gametes are the spermatozoa or sperm,

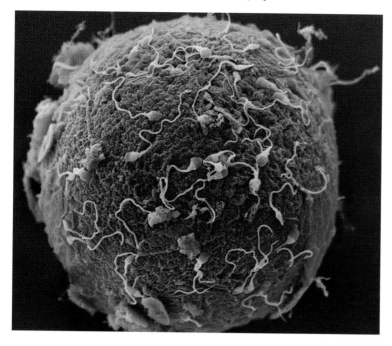

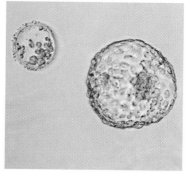

Left: Human sperm (blue) attempt to fertilize egg (red). Above: Human blastocyst, seven days after fertilization, having hatched from a glycoprotein shell (left). Top left: Sea anemone. Some organisms, such as sea anemones, can reproduce either sexually or asexually, but most animals, including humans, can only reproduce sexually. A new human life begins with the egg's fertilization. The fertilized egg becomes a blastocyst, an early-stage embryo.

produced in the sex organ called the testes; the female's are ova or eggs, produced in the ovaries.

For reproduction to happen, the sperm, which is usually smaller, must swim through a liquid and fuse with the egg. The result of that union is the zygote, or fertilized egg, which contains all the chromosomes necessary for that species, half from the mother, half from the father.

The zygote develops into an embryo, or early-stage organism, and eventually into a fully formed individual. This development happens either outside the female's body, in a structure such as a bird's egg, or inside, as in the female mammal's uterus or womb.

A PLACE FOR FERTILIZATION

Fertilization—the union of sperm and egg—can occur either inside or outside the female's body. In some organisms, from oysters to most fish, it happens outside.

This approach, external fertilization, is most common among aquatic organisms, those that live in water, where sperm and eggs are released so that the sperm can swim its way to the egg. Terrestrial, or land, organisms usually employ internal fertilization, in which the gametes unite inside of the female's body. Plants can have both external and internal fertilization.

In animals, internal fertilization requires some special adaptations. The male needs a sex organ that will place its sperm as close as possible to the egg. In mammals and some other vertebrates, this is the penis, an

European Oysters are hermaphrodites. They possess both male and female sex organs.

HERMAPHRODITES

Most multicellular organisms are differentiated into male and female sexes. The exceptions are the hermaphrodites. These naturally occurring animals, such as the earthworm, and some flatworms and segmented worms, have both male and female sex organs. For fertilization purposes, they can produce both sperm and egg; however, they generally do not use this dual generation to fertilize themselves. Instead, they use the reproductive organ that fits the situation to effect fertilization. Some oysters, for example, may begin their life as males but develop into females, while others alternate between male and female throughout their lives. Most plants are hermaphroditic, containing both stamens (male organs) and carpels (female organs).

external, fingerlike organ that hangs between the legs. When erect—stiffened by being filled with blood—it can enter the female during sexual intercourse and deposit a fluid called semen that contains sperm.

The female has a complementary, tube-shaped organ between her legs, the vagina, into which the penis fits during intercourse. The vagina provides a moist environment for sperm to swim to egg, but the egg itself is not inside the vagina or in the uterus, to which the vagina connects. Eggs travel from the ovaries to the uterus through two small channels, Fallopian tubes or oviducts. It is in one of the Fallopian tubes that fertilization usually occurs.

A cow nursing her calf. In mammals, offspring are born sufficiently weak to require continuing care from their parents.

CHAPTER 6

BREATHING AND EATING

Left: Atlantic bottle-nosed dolphin lifts itself above the water. Top: A snake sheds its skin. Bottom: Red blood cells. Respiration takes on a unique physiological function in air-breathing dolphins and other aquatic mammals. Molting, the shedding of body coverings, is a physiological function in snakes. The transporting of oxygen by red blood cells is another physiological function in animals.

The architecture of multicellular organisms is impressive, but the functioning of all those parts is even more astounding. Physiology is the study of how living things function—how they use their tissues, organs, and systems to breathe, eat, move, grow, reproduce, get rid of wastes, fight off invaders, and regulate all of those activities.

Some organisms have physiological functions that are unique to them. Only plants—and certain microorganisms equipped for the same task—can make food from sunlight through photosynthesis. Sexual intercourse happens only in mammals and other organisms in which fertilization occurs inside the female's body. Many animals molt periodically, shedding body coverings such as hair and exoskeletons. Others hibernate, entering an inactive, sleeplike state during winter. Even when organisms carry out the same functions, they may do so in different ways. To transport oxygen in the blood, mammals use a molecule called hemoglobin, while insects use hemolymph. The human stomach digests food in one chamber, the cow stomach in four. Physiology is full of variety, but it is always life in action. Biologists can study anatomy by dissecting a dead frog. Only a living organism, however, can exhibit physiology.

Solar-Powered Organisms

The most significant distinction within the biosphere may be that between organisms that can make their own food and those that cannot. Those that can make food from sunlight through the process of photosynthesis are known as phototrophs and include plants, algae, and certain bacteria. Phototrophs are generally able to live more simply than organisms that must search for and eat food; they have no digestive system, legs, nose, or teeth. Yet even for phototrophs, acquiring food is a complex business.

RAW MATERIALS

Photosynthesis is made possible by chlorophyll, a green pigment with the property of absorbing solar energy when hit by sunlight. In plants, chlorophyll is housed in chloroplasts, which are organelles in the cells of leaves.

It is chlorophyll that gives leaves their green color. In the fall, when chlorophyll production in deciduous trees cease, the underlying brilliant reds or yellows are unmasked. They were hidden by the chlorophyll.

Each leaf of a plant is a factory for making food from sunlight. And like every factory, the leaf must be supplied with raw materials. The raw materials for photosynthesis are carbon dioxide and water. Carbon dioxide in the air enters the leaves through pores called stomata. Water in the soil is absorbed by the plant's roots, and then rises through the stem through specialized tissue called xylem. Then the water passes into the leaves through veins, which are pipelines that run throughout each leaf.

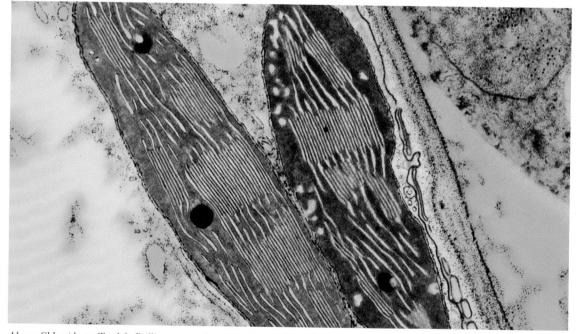

Above: Chloroplasts. Top left: Brilliant green algae. The chlorophyll inside plants and algae is what makes photosynthesis possible. In plant cells, chlorophyll is housed in structures called chloroplasts, which give leaves their green color.

PRODUCTION

When the raw materials are all in place, photosynthesis can begin. Sunlight hits the leaf, and the chlorophyll in its cells traps energy from the light. The energy splits water molecules into molecules of hydrogen and oxygen. The oxygen passes into the air through the stomata as a waste material. The hydrogen combines with carbon dioxide to produce glucose, a simple sugar that is the plant's most basic food.

The chemical formula of this carbohydrate molecule reveals that it is composed of 6 atoms of carbon (C), 12 atoms of hydrogen (H), and 6 atoms of oxygen (O): $C_6H_{12}O_6$.

The sugar produced by photosynthesis is carried to the rest of the plant in phloem, tissue composed of tubelike cells. The sugar is burned to release energy for the plant's activities, or chemically changed to form fats and starches. It may also be combined with minerals to form proteins, vitamins, and other essential substances. The minerals enter the plant dissolved in the water that the roots absorb.

THE USES OF PHOTOSYNTHESIS

The food produced by photosynthesis is the basis of life not only for plants, but also for nearly all other organisms. Glucose is one important food yielded by photosynthesis: we taste it whenever we eat such dishes as honey or grapes. But as food factories, plants use glucose to produce many other foods, from complex carbohydrates such as starch to the proteins that are the building blocks of cells. Other organisms—whether herbivores or the carnivores that eat the herbivores—ultimately rely on plants for this bounty.

Photosynthesis yields one more benefit to the global ecology: oxygen. As far as photosynthesis goes, oxygen is a waste product expelled into the air. But it is essential to the metabolism of many organisms—those, including both animals and plants, that have aerobic respiration. It is photosynthesis that provides them with the oxygen needed for life.

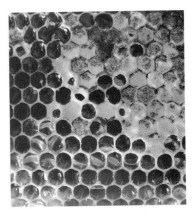

Honeycomb filled with honey. After collecting sugar from plants in the form of nectar, worker bees turn it into honey. Then they store it in honeycombs, until ready for harvest.

Venus flytrap. This plant eats insects to get nitrogen.

CARNIVOROUS PLANTS

Most plants do not attract and eat prey. But some that live in an environment without adequate nutrition have evolved to become carnivorous. One example is the Venus flytrap (*Dionaea*). This one-foot-tall plant thrives in damp surroundings such as bogs, which lack necessary nitrogen. To get it, the flytrap catches insects in its leaves and digests them.

The plant's white blossoms have two-part leaves with hinged lobes and bristles, and pods with a sticky substance that attracts insects. When the unsuspecting insect boards the appealing plant, it becomes trapped in its sticky leaves. A fluid in the leaf glands processes the insect's soft parts, and once the insect is digested, the leaf trap opens to take in further nutrition. Other carnivorous plants such as sundews (*Drosera*) and pitcher plants (families *Nepenthaceae* and *Sarraceniaceae*) also have leaves that attract and capture insects.

Obtaining Oxygen

All aerobic organisms, including plants and animals, need oxygen. Oxygen reacts with food in the body to release the energy needed for biological functioning, with carbon dioxide as one of the waste products. Oxygen abounds in the atmosphere—the air is about 21 percent oxygen—and as a dissolved gas in water. But for both land and aquatic organisms, the difficulty is in moving oxygen from the environment into their bodies and removing the carbon dioxide.

Some organisms take in oxygen through tiny holes in their body surfaces—stomata in the case of plants, spiracles in the case of insects. But vertebrates have more complex mechanisms. For land dwellers, those mechanisms are lungs, for water dwellers, gills.

LUNGS

In land-dwelling vertebrates such as reptiles, birds, and mammals, lungs are the organs that obtain oxygen. They do so by breathing, a muscular motion that rhythmically draws air into the lungs (inspiration or inhalation) and forces it out (expiration or exhalation). In mammals, breathing begins with the diaphragm, a large, dome-shaped muscle under the lungs. Attached to the rib cage, the diaphragm separates the chest cavity (containing the heart and lungs) from the abdomen (containing the stomach and intestines).

At a regular signal from the brain, the diaphragm contracts, causing its dome to flatten and the ribs to move outward. This action expands the chest cavity and increases the volume of the lungs, creating a slight vacuum; the lungs now contain less air pressure than the atmosphere outside. That vacuum pulls air into the lungs. The air flows in through the nose and mouth, continues through tubes called the pharynx and the trachea, or windpipe, and fills the lungs. The diaphragm then relaxes, snapping back into a dome. That shrinks the chest cavity and drives air back outside the body.

The air that is exhaled, however, is not the same as the air inhaled. Inhaled air is rich in oxygen, whereas exhaled air is low in oxygen but high in carbon dioxide. The exchange of gases happens inside the lung, in millions of tiny air chambers called alveoli. Each alveolus has thin, moist walls that contain networks of minute blood vessels called pulmonary capillaries. The alveoli walls are so thin that gas

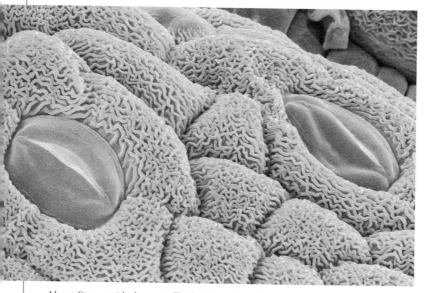

Above: Stomata (dark green). Top left: Jogger breathing heavily. Animals and plants need oxygen to survive. In humans, the lungs work harder as the need for oxygen increases during strenuous activity. In plants, oxygen enters through stomata.

molecules can flow easily through them. As blood from the rest of the body flows through the pulmonary capillaries, the blood releases carbon dioxide and picks up oxygen. From there, the bloodstream carries the oxygen to all parts of the body.

GILLS

The aquatic vertebrates called fish take in oxygen through gills, respiratory organs on each side of their heads. Water containing dissolved oxygen enters a fish's mouth, goes through the pharynx, and exits across a set of curved gill arches. Attached to the gill arches are threadlike gill filaments. As the water passes over the gill filaments, oxygen diffuses out of the water and into blood capillaries. At the same time, carbon dioxide diffuses out of the bloodstream and into the water.

Other aquatic organisms also have gills, including most crustaceans. Amphibians breathe with gills in their early, aquatic stages, although adults that live on land generally breathe with lungs.

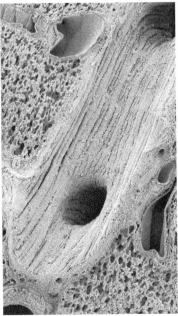

Right: Bronchiole (yellow) and alveoli (pink) in lungs. Top: A shark's gills appear just above its fins. Undersea, gills are the main organ of respiration; on land, lungs. All fish breathe through gills, but sharks have prominent gill slits because they lack a gill cover, the bony plate that hides the gills in most fish. Inside lungs, the bronchioles transport air to the alveoli, sites of gas exchange.

Processing Food

All organisms that eat food must have a process for digesting it, or breaking it into simpler molecules that their cells can use. In the typical animal, food enters through the mouth, but that is only the beginning of its journey. A complex digestive process follows, often involving the collaboration of several organs, performing its share of the labor.

CHEWING AND GRINDING

When the food is solid, chewing and grinding is often the first step of the job. Chewing insects, such as termites, grasshoppers, and beetles, use their mandibles—grinding jaws surrounding their mouths—to tear and chew solid food. But insects adapted to sucking liquid food have tubes or stylets (needles) to get their nourishment—nectar for butterflies, blood for bedbugs.

Many animals do their chewing with teeth, those bonelike structures in the mouth. In addition, a number of animals, such as birds, fish, and insects, have gizzards, muscular organs that break up and grind food.

CHEMICALS

Enzymes and other digestive chemicals are an important part of the digestive process. Enzymes are catalysts, substances that speed up a chemical reaction without being consumed by the reaction. The use of enzymes begins in the mouth, where many organisms produce the clear, sticky fluid called saliva. In many animals, including humans, saliva contains the enzyme ptyalin or amylase, which changes some complex starches into simple sugars. The act of chewing with the teeth not only breaks up food mechanically, but it also begins to break it up chemically, as saliva mixes with the food.

In vertebrates, food passes down the esophagus and into the stomach, where more enzymes go to work. The baglike stomach, which secretes gastric juice, produces a strong, churning motion that mixes food with the juice. Gastric juice includes the enzyme pepsin, which helps break down proteins into smaller molecules called polypeptides. It also includes hydrochloric acid, a highly corrosive acid that kills many microorganisms in the food and works on digesting lipids, proteins, and carbohydrates. Hydrochloric acid is so strong that it could digest the stomach itself, except for the stomach's protective measures. Chief among these measures

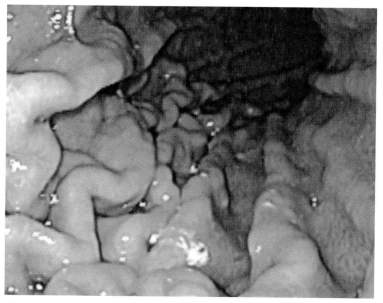

Above: Lower section of human stomach. Top left: Sheep grazing. Sheep, like other ruminants, have a four-cavity stomach. The human stomach has just one cavity.

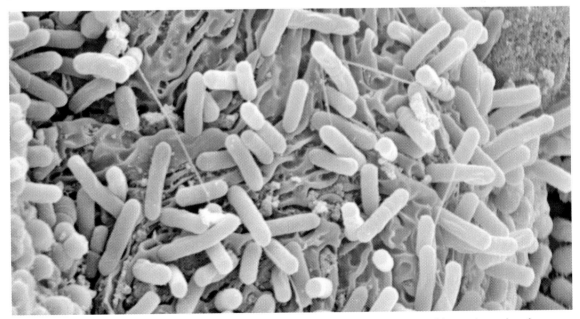

Stag beetles, like other chewing insects with mandibles, use their large external grinding jaws to hold or bite food. Among males, they serve as a weapon during fights. Mandibles are made of various invertebrate mouthparts.

include an alkaline (basic) layer of mucus along the stomach lining that neutralizes the acid.

Food progresses from the stomach into the small intestine, where further chemicals complete the process of digestion. Several organs collaborate here. The liver secretes bile into the small intestine; bile is a detergent substance that emulsifies, or breaks up, large fat globules into smaller droplets. The pancreas contributes pancreatic juice, which contains sodium bicarbonate and various enzymes specialized for digesting particular substances: amylases for starch, lipases for fat, and proteases for protein. The walls of the small intestine themselves produce intestinal juice with digestive effects. When food is completely digested, it is absorbed by blood and lymph vessels in the intestinal walls and transported into the circulatory system.

MICROORGANISMS

In many cases, microorganisms inside a larger organism cooperate in digesting food. This type of relationship, in which two different species of organisms live together in a close relationship, is called symbiosis. Bacteria in the human small intestine assist in digestion, while bacteria in the large intestine help in the processing of wastes. A ruminant—a grazing animal such as a cow, sheep, or deer—depends on bacteria and other microbes in its complex, four-cavity stomach to digest the tough plant material cellulose.

Escherichia coli bacteria in a rat's stomach. E. coli live in human and animal digestive systems. Most strains are harmless.

81

To Every Corner of the Body

Some organic processes must go on continuously, while others can be temporarily stopped. For example, if breathing stops for more than a few minutes, a person dies. But he or she can go without eating for several days and suffer nothing worse than hunger. Circulation, the transporting of nutrients, oxygen, and other vital materials throughout the body, is more like breathing than eating; it must go on all the time, or the result is death. Circulation must also reach all body tissues, or those tissues will die.

Different organisms handle circulation in different ways. Plants transport essential materials through their system of tubes called xylem and phloem. Many invertebrates, such as insects, have open circulatory systems in which blood is housed in an open cavity rather than blood vessels. Others, such as earthworms, have closed circulatory systems with blood vessels but no heart; instead the contraction of vessels pumps the blood. But for vertebrates, the heart is the beginning of circulation. It is the pump that keeps blood flowing continuously. The beating of the heart is the sound of this pump in constant action.

THE HEART

The heart beats in response to a signal from the brain, which adjusts its activity in response to need. The heart beats faster when an animal is frightened and its legs need more oxygen to help it run away; it beats slower when the animal is asleep. The heartbeat is the sound of two

Human heart. At the right of this pump is the vena cava, through which blood arrives from the body's tissues.

stages of activity in its muscle fibers: contraction (systole) and relaxation (diastole). When it relaxes, it takes in blood that is returning from circulation. When it contracts, it sends out blood to circulate through the body.

In humans, the heart consists of two pumps side by side. The left and stronger side receives oxygen-rich, or oxygenated, blood from the lungs and pumps it to refresh all of the other organs and tissues of the body. The right and weaker side receives oxygen-poor, or

Above: Earthworm, an animal without a heart.
Top left: Xylem and phloem in nasturtium leaf.

deoxygenated, blood from the body's tissues and pumps it to the lungs where it is replenished with oxygen.

The portions of the heart that receive blood, one on each side, are upper chambers called atria. Two lower chambers, one on each side, pump blood; they are the ventricles. Valves between each atrium and its associated ventricle control the flow of blood from one to the other. As the ventricles contract, pumping blood out, the valves close, preventing blood from flowing back up into the atrium. Afterward the valves open, allowing blood to flow from atrium to ventricle.

BLOOD VESSELS

Blood flows away from the heart through vessels called arteries, of which the longest and largest is the aorta. It begins at the left ventricle and branches out so that blood goes to every corner of the body delivering oxygen. The tissues take the oxygen and send back oxygen-poor blood through vessels called veins. The oxygen-rich blood is a bright red, because of the presence of oxygen; the oxygen-poor blood is a brownish-red because of its lack. There is one case in which an artery carries oxygen-poor blood and a vein oxygen-rich blood: the pulmonary artery, through which the heart sends the deoxygenated blood to the lungs, and the pulmonary vein, through which the lungs sends the blood back loaded with oxygen to continue the cycle.

Computer representation of human circulatory system. The body's main arteries (red) and veins (blue) branch out from the heart (red, upper center). The arteries carry oxygen from the heart to the body's tissues; the veins return oxygen-poor blood to the heart.

THE BLOOD OF AN INSECT

Anyone who has squashed an insect knows that it does not bleed red blood. Insects, among other invertebrates, do contain blood, but it is green, yellow, or clear, not red, because its function is not to bring oxygen to cells. Instead, insects take in oxygen through air tubes located on their sides. Further, blood does not circulate through blood vessels, as it does with mammals. In insects, blood is part of an open circulatory system that provides a suspension system for internal organs.

Honeybee. Bees, like other insects, have blood that is not red in color and does not circulate through blood vessels.

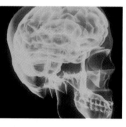

Maintaining Order

Although some organisms are simple enough not to require an elaborate control system, for vertebrates and complex invertebrates it is essential. Regulating the rate of breathing and heartbeat, coordinating the organs in vital systems, arranging legs to produce forward motion, processing sensory perceptions, responding to danger, and maintaining body temperature—all are tasks for a powerful system of control. In vertebrates, the primary control center is the brain. It exercises control through the nervous system, the body's network of nerves, and the endocrine system, the assortment of glands that secretes the chemical messengers called hormones.

THE STRUCTURE OF THE BRAIN

In humans, the brain has three main parts: the cerebrum, which makes up 85 percent of the brain by weight; the cerebellum, which lies below the back part of the cerebrum; and the brain stem, a stalk-like structure that links the brain to the spinal cord. The cerebrum is responsible for thought, memory, and the integration and processing of sensory information. The cerebellum handles balance, posture, and coordinated movement. The brain stem controls basic vital functions such as breathing and heartbeat. In addition, through its connection to the spinal cord, the brain stem links the brain to other parts of the nervous system. The brain is directly connected to certain parts of the body through other routes: for example, the two optic nerves link the brain to the eyes.

Other vertebrates also have parts corresponding to the sections of the human brain, but the human cerebrum is much more developed than in any other animal. Its greater complexity is what accounts for humanity's greater intellectual capacity.

HOW THE BRAIN WORKS

The brain operates by receiving signals from the rest of the nervous system, processing those signals, and acting upon them. Processing the signals involves sending them to the correct area of the brain. For example, if a car is hurtling toward a person, that visual information hits the person's eye, travels along the optic nerve, and reaches a part of the cerebrum called the visual cortex. Neurons in the brain analyze and interpret the information, and an appropriate

Above: Human brain. Top left: X-ray of human brain. In vertebrates the brain is the body's primary control center, and the cerebrum handles mental processes.

A lone wildebeast chased by a hungry lion. The adrenal glands, found in many animals, produce the stress hormones adrenaline (epinephrine) and noradrenaline (norepinephrine). These hormones prepare the body for stressful situations, such as taking flight from predators. The brain signals the adrenal glands to produce the hormones, which increase heart rate, breathing, and conversion of food to energy in muscles.

signal is generated—in this case, "Get out of the way." That signal travels through motor neurons, nerve cells that relay instructions to other parts of the body. In this case, the message goes to the leg muscles, which promptly get the person away from the car.

In addition, the brain signals the adrenal glands to produce the hormone adrenaline, or epinephrine, which increases heart rate, breathing, and muscular power, as a support for the leg muscles.

TRANSMITTING MESSAGES

The signals that are transmitted throughout the nervous system are called nerve impulses. Each nerve impulse involves both electricity and chemistry.

It originates when an electric charge inside a neuron (nerve cell) triggers the release of neurotransmitters, molecules that act as chemical messengers.

Once triggered, neurotransmitters cross a synapse, a small gap that separates one neuron from another. The neurotransmitters then attach themselves to sites called receptors on the surface of the other neuron. That stimulates the receiving neuron to produce

a nerve impulse of its own or to suppress impulse production. In this way, a nerve impulse can travel along the length of a nerve, a strand of tissue that contains many neurons, or be shut off, depending on the messages carried by neurotransmitters.

Dogs fighting. By preparing the body for either "fight or flight," stress hormones ready animals not just for fleeing but for chasing or fighting, as these puppies are doing.

Fighting Invasion

Defense against invasion is one task that every organism must perform. Whether composed of one cell or many, whether photosynthesizing or not, organisms need to protect themselves from infection, or attack by microorganisms. Even bacteria are in danger of being taken over by viruses—nonliving microscopic particles that hijack cell machinery to generate more viruses. The immune system of an organism wards off invaders by distinguishing self from nonself. The immune system then destroys or neutralizes anything inside the body that is nonself.

IMMUNE SYSTEMS

Immune systems in multicellular organisms vary in complexity. Some plants have waxy secretions (cuticle) that help hold in moisture and keep out microbes. Their fruits are high in vitamin C and bioflavonoids, which are antibacterial and antiviral. But plants do not have the more elaborate immune defenses that animals possess.

Central to the immune system of humans and many other animals is the white blood cell or leukocyte. A component of blood, this round, colorless type of cell is specialized to identify and fight invading microbes and harmful substances. Some kill bacteria by engulfing and digesting them. Others produce antibodies, proteins that make microscopic invaders harmless. Antibodies do this by binding with antigens, substances that the body regards as foreign, and preparing them for destruction by white blood cells. The lymphatic system, the enter the bloodstream. The lymph nodes also produce lymphocytes, a type of white blood cell. The tonsils, spleen, and thymus also contain lymphoid tissue (resembling the tissue of the lymph nodes) that produces lymphocytes and aids the body in the fight against infection.

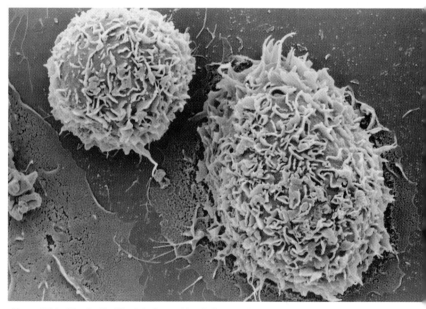

Above: White blood cells. Top left: Insect bite. Inflammation is one way the body responds to invasion by foreign particles, such as those introduced by an insect bite. As blood vessels expand in the area, white blood cells are attracted.

network of vessels that carries fluids from body tissues to the bloodstream, is a principal part of the immune system. It contains lymph nodes, masses of tissue that filter out bacteria and other foreign particles so they do not

THE DRAWBACKS OF IMMUNITY

The body's immune response is not always pleasant. It can present itself as inflammation, a condition that tissues undergo in response to injury, infection, or irritation.

Blood vessels expand, allowing more blood to flow into the area and attracting white blood cells. The result can be destruction of invaders but also swelling, redness, heat, and pain, as the expanding tissues press on sensory nerves.

The immune system can also malfunction. In autoimmune diseases, such as systemic lupus erythematosus, the immune system mistakes its own tissues as foreign and attacks them, causing serious damage. In common allergies, the immune system overreacts to a usually harmless foreign substance, such as pollen or dust, and launches unnecessary immune reactions, such as sneezing, coughing, or itching.

In other cases, the immune system needs a boost from medical science. Vaccines are an example. For the body to produce antibodies to particular bacteria or viruses, it needs to have been exposed to those microbes. But such exposure can mean a fatal or disabling case of diphtheria or polio. A vaccine for a given disease introduces killed or

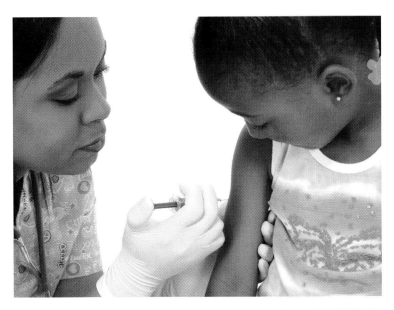

weakened microbes that trigger production of antibodies without causing disease. The continuing presence of those antibodies is what immunizes the person against that illness.

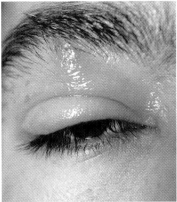

Right: Eyelid swells following an allergic reaction to a mosquito bite. Above: Girl is immunized. The immune system can be boosted by vaccines that prompt the body to produce antibodies to particular bacteria or viruses. Sometimes the system overreacts, as in allergic responses to common irritants.

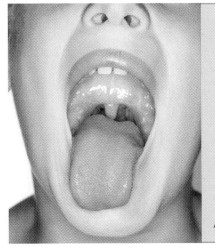

TONSIL OVERLOAD

Tonsils are almond-shaped lymphoid tissue masses at the back of the human throat. They are thought to defend the body from infection and are part of the immune system, but they have their limits.

Present in higher primates, tonsils contain white blood cells that consume foreign invaders such as airborne bacteria and viruses. In humans, the tonsils include palatine or faucical tonsils near the palate, pharyngeal tonsils or adenoids, and two lingual tonsils at the base of the tongue. The tonsils aid in fighting infection, yet sometimes, they become so inflamed with the disease called tonsillitis they have to be removed. Why? They are overwhelmed by a large number of bacteria or viruses. Like a computer, the tonsils "crash."

Boy shows his tonsils. Tonsils can become inflamed with tonsilitis.

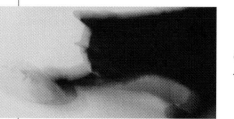

Sensation and Motion

When a cat sees a mouse and pounces, two factors determine whether the cat will get a meal: how well it sees and how well it pounces. The function of sense organs, such as the cat's eyes, is to allow the organism to receive information from its environment, such as where the mouse is. And one function of muscles, including those in the cat's legs, is to allow it to do something useful with the information, such as catch the mouse.

Organisms have many methods for sensing and moving. The hydra, a small aquatic animal, has sensory cells scattered around its surface. The one-celled amoeba moves by pushing out its membrane to form a fingerlike structure called a pseudopod (false foot), into which the rest of the organism flows. But the senses of more complex animals, such as vertebrates, are capable of relaying much more information about the environment. And the musculoskeletal system of vertebrates, in which muscles are attached to the skeleton, is particularly adept at moving large land animals.

THE SENSES

Senses come in two varieties: external senses, which receive information about the world outside the body, and internal senses, which receive information about the interior of the body. The well-known five senses—sight, hearing, smell, taste, and touch—are the major external senses of the human being. Other animals have these too, to greater or lesser degrees. For example, a dog can smell and hear far better than a human, although it lacks the human's full-color vision. Some animals

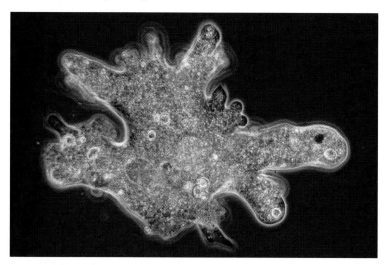

Above: Amoeba. Top left: Cat pouncing. To catch a mouse, a cat needs sharp senses, strong muscles, and exquisite coordination between them. Even on the unicellular level, motion is critical to survival. Amoebas move by extending their fingerlike pseudopods.

Right: German shepherd. Bred from its hunting ancestors, the domestic dog has senses of hearing and smell that are far superior to human senses. The canine lacks full-color vision, but is sensitive to motion and can see well in dim light.

have external senses that humans lack. Certain birds and insects are thought to be able to detect the direction of Earth's magnetic field. The rattlesnake has pit organs on its face that enable it to find warm-blooded prey by sensing the prey's body heat.

Internal senses include those that contribute to feelings of hunger, thirst, fatigue, and pain. They respond to stimuli in the body's various systems, such as the digestive, circulatory, respiratory, and nervous systems.

Senses depend on receptors, which are cells specialized to detect a certain kind of stimulus. When stimulated, these receptors initiate transmission of a signal along sensory nerves, so that they can reach the central nervous system and be acted upon. For example, light-sensitive cells in the retina, a part of the eye, absorb light rays and convert them into electrical signals that travel along the optic nerve to the brain. Receptors that respond to chemical changes inside the body are called internal chemoreceptors.

THE MUSCULOSKELETAL SYSTEM

The ability of muscle cells to contract is what makes muscle tissue useful for motion. When electrically stimulated by a nerve, the cells in a muscle contract, producing motion. The motion is amplified in animals with internal skeletons, such as humans. There the

The cheetah, the fastest land animal, can reach speeds of up to 60 miles per hour.

WHY IS THE CHEETAH FAST?

A cheetah is the fastest land mammal across short distances. This large cat, found on the scrubby plains of eastern and southern Africa, can run at speeds of up to 60 miles per hour. Its superb running ability is rooted in its anatomy and physiology. It has outstretched claws (rather than retractable ones) and ridged foot pads that provide traction on sprints. Its head, which is small in proportion to its body, reduces wind resistance, and its legs are long for its torso. Its spine acts like a giant spring, and its oversized heart, lungs, and liver deliver oxygen and energy efficiently. Like a sports car, the cheetah is built for speed.

contracting tissue of the muscle pulls on a bone as if it were a lever, with an attached joint as a fulcrum. That lever applies force for such motions as getting up, walking forward, lifting things, or pouncing on a mouse.

Muscles are particularly valuable for motion because they adapt to how they are used. For example, they grow larger and stronger with use. That is what makes possible the sport of bodybuilding, in which the repeated lifting of weights forces the athlete's muscles to grow larger.

Bodybuilders in a flexing pose. Bodybuilding is possible because muscle tissue grows larger and stronger with use.

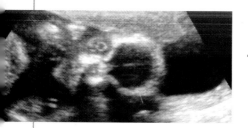

The Next Generation

A sperm cell from a male and an egg cell from a female unite to form a fertilized egg, which develops into a new individual. That much is the same in all species that reproduce sexually, from flowers to gorillas. But there is wide disparity in the physiological details of reproduction and growth.

REPRODUCTION

Many species practice elaborate courtship rituals, behaviors by

which they find and attract mates. Male frogs call and male birds sing to attract females of their species, as well as to warn rival males to stay out of their territory. The sounds allow females to identify the sound-makers as males of their species. Male fireflies use light-emitting organs to flash a light signal to female fireflies, which respond with a light of their own. Other kinds of displays may attract a mate.

The peacock, or male peafowl, spreads its magnificent train of greenish feathers into a splendid fan as it parades in front of the female. A bull, or male moose, will butt antlers with another bull to see which one will win the cow.

Once a mate is found, fertilization can take place. Depending on the species, it may happen through sexual intercourse, as the male inserts part of his body inside the female's body, depositing sperm so it can unite with the egg. Or it may happen outside the female's body, as when fish deposit sperm and eggs in the water, so they can unite there.

Left: Peacock parading its train of feathers. Top left: Ultrasound of fetus in its mother's womb. Sexual reproduction begins with finding and attracting mates, sometimes with displays such as the peacock's feathers. After fertilization and development, it ends with a fully formed new life.

DEVELOPMENT

A long process of development is needed to transform a single fertilized egg cell, or zygote, into a fully formed plant or animal. First, the zygote goes through a series of divisions that generates many cells. As the cells of this early-stage organism, known as an embryo, continue to divide, they differentiate into the different tissues and organs proper to that plant or animal. The differentiation is directed by genes inside the organism and also by the mother's genes. The genes direct the development through the production of chemicals called morphogens. These determine that one part of a fruit fly, for example, will become the head, another the anus, and that genes in certain regions of the embryo will be switched on so that the regions become legs, not antennae or wings.

The development of the embryo may take place inside the body of the mother, as in the wombs of mammals, or outside of it, as in the eggs of birds. That environment is generally more protective than the outside world will be. Food is provided, through the maternal blood supply in the womb or the yolk in the eggs. Temperature control is provided, through the mammalian mother's internal body heat or the bird sitting on its eggs.

Then, sufficiently developed to survive in the open, the new individual breaks out. It is born as contractions in the mother's womb push it out of the mother's body, or hatched as it breaks through the eggshell. At that point, the offspring of most species usually have to undergo further stages of development.

Insects and many aquatic animals have larval stages, in which their early forms differ radically from those of adults. Only later, through the process of metamorphosis, do they become adults, turning from tadpoles into frogs and from caterpillars into butterflies. Humans have their own stages of development, from child to adolescent to adult. Through such processes, the next generation of each species matures.

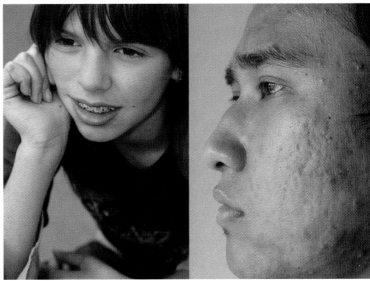

Top: Pregnant Shetland pony. Above left and right: The changes of adolescence—braces (left) and acne (right). Between the conception of a new individual and the emergence of a new adult, a long process of development intervenes. In mammals such as horses, much of the development is completed inside the mother's womb. Human offspring, as in many other species, undergo further stages of development. Their development from childhood to maturity is particularly long, and is accompanied by the sometimes traumatic changes of adolescence.

Famous Biologists

Mary Anning
(1799–1847) British paleontologist. Self-taught, she discovered many fossils, including those of an ichthyosaur, a pterodactyl, and the first known plesiosaur.

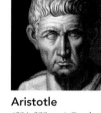

Aristotle
(384–322 BCE) Greek philosopher and scientist whose views on zoology, botany, and other areas of biology became authoritative in Europe during the Middle Ages.

Claude Bernard
(1813–78) French physiologist, founder of experimental physiology. Made discoveries about the operation of the liver, pancreas, and nervous system.

Stanley Cohen
(b. 1922) American biochemist who, with American biochemist Herbert Boyer, developed genetic engineering through the artificial recombination of genes.

Francis Crick
(1916–2004) British biologist who, with James Watson and Maurice Wilkins, defined the structure of DNA, the molecule that transmits genetic information.

Georges Cuvier
(1769–1832) French naturalist who studied comparative anatomy. Helped establish paleontology by including prehistoric animals in his comparisons.

Charles Darwin
(1809–82) British naturalist, formulator of the theory of evolution by natural selection. Evolution is the central, unifying tenet of all the biological sciences.

Rosalind Franklin
(1920–58) British chemist and molecular biologist. Her X-ray studies of DNA contributed significantly to the elucidation of the structure of that molecule.

Galen
(c. 130–c. 200) Greek physician in Rome. He greatly contributed to anatomy and physiology. Europeans considered his views authoritative into the Renaissance.

William Harvey
(1578–1657) British physician who discovered how blood circulates in mammals. He established that the heart is a muscle that works like a pump.

Hippocrates
(c. 460–c. 370 BCE) Greek physician regarded as the father of medicine. He taught that diseases had natural causes and emphasized personal observations.

Robert Hooke
(1635–1703) English scientist. Using a microscope, he discovered plant cells. He published the first drawings of cells in his 1665 book *Micrographia*.

Robert Koch
(1843–1910) German physician. Showed that specific bacteria cause specific diseases. Discovered cause of TB and introduced technique for culturing bacteria.

Karl Landsteiner
(1868–1943) Austrian-born American pathologist. Identified major types of human blood. With Philip Levine and Alexander Wiener, discovered the Rh factor.

Antoine Lavoisier
(1743–94) French chemist who applied the techniques of chemistry to physiology. He described the role of oxygen in the respiration of animals and plants.

Mary Leakey
(1913–96) British archaeologist and anthropologist. With husband Louis Leakey, discovered evidence of australopithecines and other human ancestors in Africa.

Anton van Leeuwenhoek
(1632–1723) Dutch scientist. Developed microscopes that allowed him to first clearly describe bacteria and other microscopic life.

Carolus Linnaeus
(1707–78) Swedish naturalist and botanist. An authority on plants, he established the modern scientific method of classifying organisms and naming species.

Barbara McClintock
(1902–92) American geneticist. Discovered mobile genetic elements, genes that can change their position on chromosomes.

Marcello Malpighi
(1628–94) Italian anatomist. Using a microscope, he was the first to observe blood circulation through the capillaries. Made key discoveries about human anatomy.

Gregor Mendel
(1822–84) Austrian botanist who formulated the fundamental laws of heredity. He discovered that traits are transmitted by basic hereditary units later called genes.

Thomas Hunt Morgan
(1866–1945) American geneticist. Using fruit flies for experiments, discovered that genes are located on chromosomes in a fixed linear order.

Louis Pasteur
(1822–95) French chemist who contributed to the germ theory of disease. He developed pasteurization and invented vaccines for anthrax and rabies.

Matthias Schleiden
(1804–81) German botanist. With physiologist Theodor Schwann, he developed the cell theory, which states that the cell is the basic unit of life.

Andreas Vesalius
(1514–64) Flemish anatomist regarded as the founder of human anatomy. Based on his own dissections of cadavers, he made many anatomical discoveries.

Leonardo da Vinci
(1452–1519) Italian artist who advanced the study of human anatomy with hundreds of anatomical drawings based on dissections of human cadavers.

Rudolf Virchow
(1821–1902) German pathologist who argued that disease results from disturbances in cell function. He was a founder of the discipline of cellular pathology.

James Watson
(b. 1928) American biologist who, with Francis Crick, discovered the structure of DNA. He was later one of the architects of the Human Genome Project.

Ian Wilmut
(b. 1944) British scientist who, in 1996, was the first to clone a mammal, a Finn Dorset lamb named Dolly, from an adult cell of another animal.

Edward O. Wilson
(b. 1929) American entomologist and sociobiologist. He is the founder of sociobiology, which applies evolutionary principles to the study of social behavior.

Discoveries in Biology

c. 28,000 BCE
Cro-Magnons in France and Spain create cave paintings of bears, bison, cattle, horses, and mammoths that reflect close observation of nature.

c. 12,000 BCE
Dogs are domesticated from wolves in Mesopotamia (present-day Iraq).

c. 8,000 BCE
The practice of agriculture begins in northern Mesopotamia. It will spread over the next two millennia into Mexico, China, and Japan.

Mesopotamian bas-relief

c. 2,800 BCE
The *Shennong Herb-Root Classic*, a collection of writings on medicinal plants, is attributed to Chinese emperor Shennong. His work marks the development of early biological knowledge in China.

c. 500 BCE
Greek physician Alcmaeon dissects human cadavers to study them.

c. 400 BCE
Greek physician Hippocrates teaches that diseases have natural causes and that the treatment of diseases benefits from bedside (personal) observation.

c. 375 BCE
Hippocrates develops the disease theory of the four humors. The theory suggests that disease arises from an imbalance of bodily fluids called humors: blood, phlegm, black bile, and yellow bile.

c. 350 BCE
Greek philosopher Aristotle prefigures empirical scientific study as he uses "sense experience" to study aging, animals, memory, plants, and sleep.

Aristotle

c. 300 BCE
Theophrastus, Greek scientist and student of Aristotle, describes more than 500 plant species. He will be known as the father of botany.

c. 180 BCE
Greek physician Galen dissects animals and makes anatomical discoveries. His views dominated European medicine for over a thousand years.

c. 1500 CE
Italian artist Leonardo da Vinci advances the scientific study of anatomy as he creates anatomical drawings from dissections of human cadavers.

Da Vinci's drawing

1543
Flemish anatomist Andreas Vesalius publishes a systemic treatise of human anatomy, *On the Structure of the Human Body*. He is considered the founder of the study of human anatomy.

Vesalius's anatomical drawing

c. 1500–1700
The period known as the Scientific Revolution. Scholars use experimentation and quantification to make discoveries in nature, including living beings and chemicals.

1625
French philosopher René Descartes defines the concept of reflex action.

1628
British physician William Harvey shows that the heart is a pump-like muscle that produces regular contractions. His research reveals how blood circulates in mammals.

1665
English scientist Robert Hooke discovers plant cells and publishes *Micrographia*, which records cell drawings.

Hooke's drawing of flea

1668
Through a series of experiments, Italian scientist Francesco Redi disproves spontaneous generation. He shows that life cannot emanate from nonlife.

1674
Dutch scientist Anton van Leeuwenhoek is the first to describe red blood cells correctly.

Anton van Leeuwenhoek's drawing of microscopic spermatozoa.

1694
German botanist Rudolph Jakob Camerarius demonstrates that plants reproduce sexually.

c. 1750
Swedish naturalist Carolus Linnaeus establishes the modern scientific system for classifying organisms and naming species.

Carolus Linnaeus

c. 1780
English physician Edward Jenner develops a vaccine for smallpox.

Edward Jenner

c. 1790
French naturalist Georges Cuvier founds comparative anatomy, following studies of animal body types. Cuvier uses fossilized prehistoric remains, hastening the development of paleontology.

c. 1790
French chemist Antoine Lavoisier applies chemical techniques to physiology; recognizes and names oxygen.

c. 1800
French scientist Jean Baptiste Lamarck coins the word "biology" by joining two Greek words: *bios* (life) and *logos* (study).

1828
German chemist Friedrich Wohler is the first person to make an organic substance from an inorganic substance, creating urea, a product of mammal urine.

1831
Scottish naturalist Robert Brown discovers that the nucleus is a regular structure in all plant cells.

1834
French chemist Anselme Payen discovers cellulose, the main constituent of plant cell walls.

1835
German botanist Matthias Schleiden and physiologist Theodor Schwann propose the cell theory, which posits that the cell is the basic structural and functional unit of life.

c. 1840
British paleontologist Mary Anning discovers fossils of species including the ichthyosaur, pterodactyl, and plesiosaur.

1847
Hungarian physicist Ignaz Semmelweis posits the necessity of cleanliness in preventing disease, showing that the spread of childbed fever can be prevented by having doctors wash their hands.

Ignaz Semmelweis

1852
German physicist Hermann von Helmholtz applies physical principles to determine the speed at which messages travel along nerves.

Discoveries in Biology (continued)

Rudolf Virchow

1858

German pathologist Rudolf Virchow proposes that disease results from a disturbance of the cell function.

1859

British naturalist Charles Darwin publishes *The Origin of Species*, in which he argues that evolution occurs by natural selection.

Title page from Darwin's The Origin of Species

c. 1860

French physiologist Claude Bernard shows that body temperature of warm-blooded animals resides in their nervous system. He is considered a founder of modern experimental physiology.

1860s

The germ theory of disease is formulated by French chemist Louis Pasteur and others.

Gregor Mendel

1865

British surgeon Joseph Lister introduces the use of germicide (carbolic acid) on surgical instruments and surgeon's hands. It reduces surgical mortality by more than 50 percent.

1866

Austrian botanist and monk Gregor Mendel discovers that heredity units (later called genes) transmit the inherited traits of organisms.

c. 1870

French chemist Louis Pasteur develops vaccines for anthrax and rabies.

1882

German physician Robert Koch discovers the bacterium that causes tuberculosis. By showing that it and other bacteria cause certain diseases, he promotes the science of bacteriology.

1898

Dutch botanist Martinus Beijerinck discovers a filterable virus, a disease-causing particle so small it can elude a filter meant to keep out bacteria.

1900

The major types of human blood (O, A, B, and AB) are discovered by Austrian-born American pathologist Karl Landsteiner.

1902

American geneticist Walter S. Sutton suggests that chromosomes might contain the genetic factors predicted by Gregor Mendel.

1909

American geneticist Thomas Hunt Morgan discovers that genes are located on chromosomes in a fixed linear order.

c. 1930

Russian physiologist Ivan P. Pavlov demonstrates that a conditioned reflex can be generated.

1931

American geneticist Barbara McClintock shows that chromosomes can break and exchange parts during the formation of eggs or sperm.

1944

Canadian-born American bacteriologist Oswald T. Avery finds that DNA alone determines heredity.

1952

American chemists Stanley Miller and Harold C. Urey demonstrate that amino acids could have arisen naturally from simpler chemicals.

1953

American biologist James Watson and British molecular biologist Francis Crick discover the double-helix structure of the DNA molecule. It unlocks the genetic code governing heredity.

1955

American microbiologist Jonas Salk develops a polio vaccine.

Jonas Salk

1959

In Africa, British anthropologist Mary Leakey, in collaboration with husband and colleague Louis Leakey (1903–72), discovers fossil remains of the hominid *Zinjanthropus* (now called *Australopithecus boisei*).

Louis and Mary Leakey

1968

American paleontologist Robert Bakker proposes that dinosaurs may have been warm-blooded.

Konrad Lorenz

c. 1970

Austrian-born German zoologist Konrad Lorenz studies birds and describes the process of imprinting.

1973

American entomologist Edward O. Wilson founds sociobiology, which applies evolutionary principles to the study of animal social behavior, including humans.

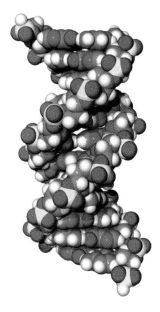

DNA molecule

1973

American biochemists Stanley Cohen and Herbert Boyer achieve genetic engineering through artificial recombination of genes.

1996

British scientist Ian Wilmut clones the first mammal, a sheep named Dolly.

2003

The Human Genome Project is completed, determining the sequence of the entire human genetic inheritance.

Human Genome Project lab

Life through Time

Life has a history, and that history is recorded in two places: rocks and bodies. As organisms died, they left behind remains, some of which became fossils buried in rocks. Those rocks have been dated and classified into a succession of geologic intervals. The fossils in the rocks give an indication of how life evolved. For example, rocks from the Permian period (290 to 248 million years ago) contain no dinosaur fossils, whereas rocks from the next period—the Triassic (248 to 206 million years ago)—do. The geologic time scale shows the order of the geologic intervals.

The history of life is also visible in the living bodies of modern organisms. The feathers that all birds share attest to their descent from a common ancestor, just as the exoskeleton of crabs and bees testifies to their relatedness. This anatomical relativeness has allowed scientists to form an evolutionary tree of life.

	Eon	Era
		Cenozoic
	Phanerozoic	Mesozoic
		Paleozoic
	Proterozoic	
	Archean	
	Hadean/Priscoan	

Geologic Time Scale

The geologic time scale presents the history of Earth in units of time. The oldest units are at the bottom, the most recent at the top. The units are nested hierarchically, with the largest-scale unit being the eon, followed by the era, then the period, then the epoch. For example, the Eocene is an epoch within the Tertiary period of the Cenozoic era, which is part of the Phanerozoic eon. The number in the right column shows how many millions of years ago a given unit of time began.

Therapods, three-toed dinosaurs believed to be ancestors of birds, first appeared in the late Triassic period.

The cycad is a tropical plant dating from the Permian period. It flourished in the Jurassic, sometimes called "the age of cycads."

Earth was formed during the Precambrian, when materials already orbiting around the sun came together to form a planet.

Based on data from the Geological Society of America

Period	Epoch	Began (millions of years ago)
Quaternary	Holocene	0.01
	Pleistocene	1.8
Tertiary	Pliocene	5.3
	Miocene	23.8
	Oligocene	33.7
	Eocene	54.8
	Paleocene	65
Cretaceous		144
Jurassic		206
Triassic		248
Permian		290
Carboniferous/ Pennsylvanian		323
Carboniferous/ Mississippian		354
Devonian		417
Silurian		443
Ordovician		490
Cambrian		543
		2500
		3800
		4600

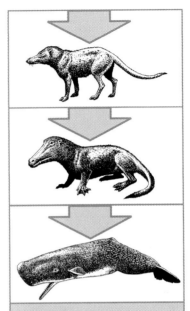

One of the branches of the tree of life leads from Pakicetus *(top), a terrestrial ungulant mammal of the Eocene, to* Ambulocetus, *(middle), whose name means "walking whale." The branch then leads to modern whales, including the sperm whale (bottom).*

EVOLUTIONARY TREE OF LIFE

All organisms that have ever lived are related, and the evolutionary tree of life shows how. The origin of life is at the base of the tree; each branching of the tree represents the evolution of a major new kind of organism. Time passes as the tree grows higher. Prokaryotes (bacteria) appear first, with eukaryotes later branching off. The eukaryotes in turn branch into protists, plants, animals, and fungi. Each of these kingdoms spawns its own branches. Some branches die off; for example, trilobites went extinct without leaving descendants.

Biomes of the World

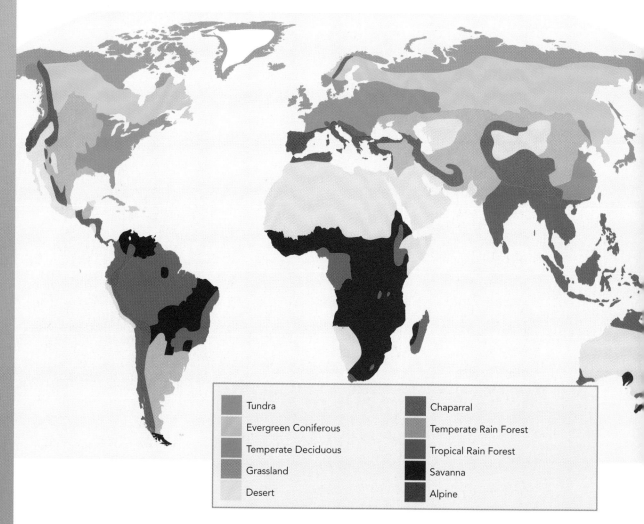

Tundra		Chaparral	
Evergreen Coniferous		Temperate Rain Forest	
Temperate Deciduous		Tropical Rain Forest	
Grassland		Savanna	
Desert		Alpine	

BIOMES DEFINED

Biologists divide the world into biomes, large geographic regions distinguished by their climate and dominant vegetation and animals. Each of these ecological zones is occupied by distinct forms of life—for example, maple trees and robins in temperate deciduous forests, and palm trees and monkeys in tropical forests. Each biome appears in many parts of the world. For example, the North American plains, the pampas of South America, and the steppes of Central Asia are all examples of the grassland biome. Climate plays an important role in determining which part of the world will exhibit which biome. Portions of formerly frigid tundra have been turning to coniferous forest as global climate warms.

Expert opinions differ as to how to classify the world's biomes. But the major land biomes are generally thought to be tundra, evergreen conifeous forest (or taiga), temperate deciduous forest, temperate rain forest, temperate grassland, desert, chaparral, tropical rain forest, tropical savanna and woodland, and alpine.

Tundra

North West Territory, Canada. The coldest of the biomes, arctic tundra regions are marked by permafrost, or permanently frozen soil. Shrubs, lichens, and flowering plants grow in the tundra.

Evergreen Coniferous

Washington State, U.S.A. Also known as taiga, or boreal forests, this large biome covers much of northern Europe and Asia, and northern North America.

Temperate Deciduous

Jiuzhaigou, China. Four distinct seasons mark this biome, which appears in North America and Europe, but also in eastern Asia. Fall color is its most pronounced feature.

Grassland

Grasslands on three continents illustrate the features of this biome. Photos show central Asia (left), North America (center), and the Argentine pampas of South America (right).

Desert

The Sahara Desert, Morocco. The world's most arid biome, deserts are inhabited by richly diverse plants and animals, all adapted to these regions' low rainfall.

Chaparral

Pinnacles National Monument, California. Drought-resistant shrubs and dwarf evergreen oaks are characteristic of this biome. It covers Mediterranean climates in the Americas, Australia, South Africa, and Europe.

Rain Forest

Divided into temperate and tropical rain forests, these biomes receive heavy annual rainfall. A temperate rain forest is at left; a tropical rain forest at right.

Savanna

Drakensberg, South Africa. Grassy hills with scattered trees and shrubs dominate this relatively arid biome. Summers are short and rainy; winters are long and dry.

Alpine

Mt. Whitney, California, U.S.A. The alpine biome contains a range of animal and plant life, which varies with altitude. Higher elevations sustain fewer plant and animal species.

Kingdoms of Life

Biologists commonly classify living things into five kingdoms, or large-scale units. The kingdoms are prokaryotes, protists, fungi, plants, and animals. This division has not gone undisputed: Some argue that there are six kingdoms, with prokaryotes divided into two distinct kingdoms, archaea and eubacteria.

Overall, the division into kingdoms, and into the finer categories of phylum, class, order, family, genus, and species, has proved useful. It offers a guide to distinguishing living things—for example, noting that spiders and flies are from the same phylum (arthropods) but different classes (respectively, arachnids and insects). It also helps in understanding their evolutionary relationships. It is not chance that puts oaks and cacti in the same kingdom of plants, but the fact that the two are more closely related to each other than either is to any animal, such as a spider or fly.

Prokaryotes	Protists	Fungi

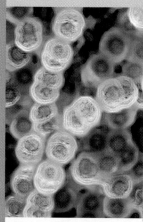

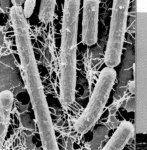

The prokaryotes differ from organisms of all other kingdoms, known as eukaryotes, in that their cells contain no nucleus. Most prokaryotes are bacteria. Streptococcus (above) is a round bacterium that is often harmful to humans; the diseases it causes include scarlet fever, meningitis, angina, and pneumonia.

Protists are largely defined as single-cell organisms that contain a nucleus. They include plant-like organisms like algae (above), that use photosynthesis.

Other protists, like the hypotrich (above), a ciliate protozoan, are more animal-like.

Fungi are eukaryotic organisms that digest their food externally. Since they do not use photosynthesis, they are not related to plants. Some commonly known fungi include mushrooms (above), yeast, and mold.

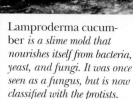

Lactobacillus acidophilus, *the bacterium in yogurt, is beneficial to humans; it converts lactose, or milk sugar, into lactic acid.*

Lamproderma cucumber *is a slime mold that nourishes itself from bacteria, yeast, and fungi. It was once seen as a fungus, but is now classified with the protists.*

Mold on an orange. Fungi serve many purposes; here it is a decomposer. Some are parasitic; Dutch elm disease, a tree parasite, and athlete's foot are both examples of parasitic fungi.

Plants	Animals

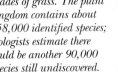

Blades of grass. The plant kingdom contains about 258,000 identified species; biologists estimate there could be another 90,000 species still undiscovered.

The red-gilled nudibranch, a sea slug that breathes through a bushy plume of extremities on its back, is classified as a mollusk.

Succulent plants like saguaro cactus in Arizona (above) and deciduous trees like the oak (below) are more closely related to each other than to any animal.

Atlantic puffins are seabirds that live along the coasts of Europe, Iceland, as well as eastern North America.

The donkey belongs to the same genus as the horse; donkeys were domesticated as long ago as 4000 BCE.

ARISTOTLE'S LOOKALIKES

Informed by the theory of evolution, modern taxonomy groups organisms by shared ancestry. In the past, however, organisms were often classified merely by physical resemblance, without regard to origin. For example, when the Greek philosopher Aristotle (384–322 BCE) studied the appearance of plants and animals, he grouped animals with tusks together, and plants with similar leaf shapes. Below are organisms from three different kingdoms that appear to have similar characteristics, with spiny or spiky protrusions emanating from a circular center. However, the cactus (bottom) belongs to the kingdom Plantae, the sea anemone (middle) to the kingdom Animalia, and the mushroom (top) to the kingdom Fungi.

From Cells to Organisms

The surface of a typical animal or plant conceals a very complex interior. On the microscopic scale, such an organism is composed of a multitude of cells, each one a factory taking apart molecules and assembling chemical products. On a larger scale, each organism is composed of organ systems, each one performing a vital function such as respiration, circulation, or digestion. The organ systems are composed of tissues, which in turn are composed of individual cells. In many organisms, a hard framework supports the entire organism, such as the ex-ternal skeleton of insects or the internal skeleton of vertebrates.

The images on these two pages display three views of the complexity of organisms: (1) the parts of an animal cell, such as those that make up human tissues; (2) the skeleton that provides the supporting framework for a human being; and (3) the major organ systems of a human being.

Parts of a Cell

Although eukaryotic cells vary in size, shape, and function, they have many similarities. All have an outer plasma mem-brane that encloses a jellylike material called cytoplasm. Within that cytoplasm are structures called organelles that do the cell's work. In eukaryotic cells, such as those of plants and animals, a central organelle, called the nucleus, directs the cell's activities. This illustration is a generalized representation of an animal cell, showing its membrane, nucleus, and such organelles as the mitochondrion, ribosome, and lysosome. Plant cells have additional structures that are not shown here, such as chloroplasts, the site of photosynthesis.

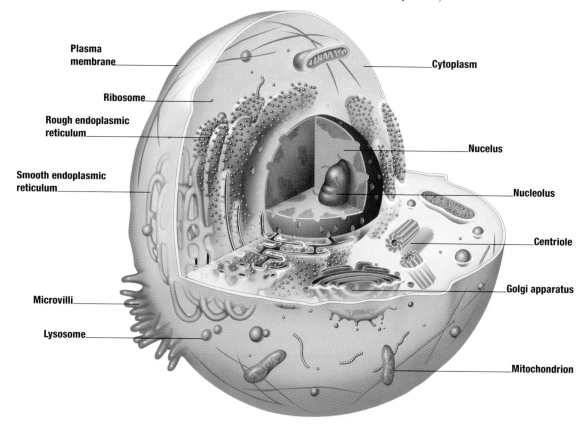

Plasma membrane
Ribosome
Rough endoplasmic reticulum
Smooth endoplasmic reticulum
Microvilli
Lysosome
Cytoplasm
Nucelus
Nucleolus
Centriole
Golgi apparatus
Mitochondrion

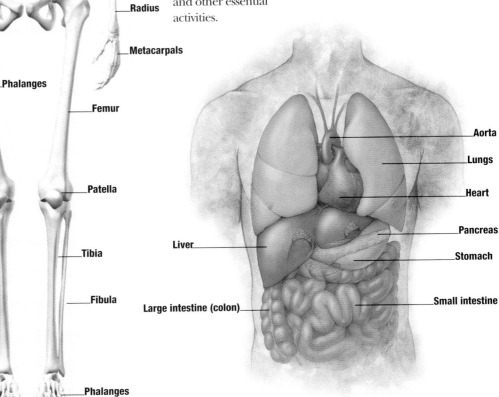

Skull

Sternum

Humerus

Ribs

Vertebrae

Pelvis

Ulna

Radius

Metacarpals

Phalanges

Femur

Patella

Tibia

Fibula

Phalanges

The Visible Human

A human being is a combination of hard and soft parts. The hard parts are the skeleton, a supporting framework of 206 bones connected to each other by joints. Some joints are immovable, as in the skull. Other joints such as those in the arms and legs are capable of movement. The human skeleton has two main parts, the axial skeleton (bones of the head, neck, and trunk) and the appendicular skeleton (bones of the arms, legs, and their supports).

The soft parts are largely organ systems, networks of organs that carry out vital functions. These include systems dedicated to digestion, respiration, circulation, reproduction, and other essential activities.

Above: Microscopic image of a kangaroo rat cell. The nucleus, containing the DNA, is shown in orange. Far left: A human female skeleton, which has a slightly wider pelvis and narrower shoulders than the skeleton of a male. Below: The major organs of a human, including the lungs, heart, stomach, and intestines. These soft parts are largely protected by the axial skeleton of the trunk.

Aorta

Lungs

Heart

Pancreas

Stomach

Small intestine

Liver

Large intestine (colon)

CHAPTER 7

THE COMMUNITY OF LIFE

Every day, a dynamic interplay between living and nonliving things takes place. Sunlight shines, rain falls, wind blows, trees grow, insects eat, mice scurry, birds fly, and humans go about their lives. These and many other interactions between living things and the environment are known as ecology. Ecologists, those who study ecology, constantly question the world around them. Where are living things found? How many living things exist? How diverse is life? The answers to these questions are crucial to the survival of the living things with which we share the planet. And, because we humans depend on ecology for our own survival, the answers to these questions are of vital importance to us.

All living things on Earth are connected—by water, air, and other materials that circulate from one part of the globe to another. But, because Earth's environments vary so vastly, certain living things thrive in particular regions and not in others. Each species has its own ecological role to play.

Relationships between living and nonliving things are not static. When resources are abundant, populations expand. When natural disasters strike, space for new life is created. Life constantly responds and adjusts to change.

Left: The floor of this temperate rain forest teems with only a few species of plants. Mosses, lichens, and ferns cover tree stumps and rocks. These plants are known as nature's first colonizers. They allow soil to accumulate, providing a resource for other plants to move in and crowd out the original settlers. Top: Sunlight breaks through the cloud canopy. Bottom: A lion prowls in the tall grass of an African savanna, one of three types of grassland. The others include steppes and prairies.

Complex Interactions

Life is full of complex interactions. For example, during a typical day in Yellowstone National Park, a pocket gopher might dig a tunnel, eat grasses, excrete waste, communicate with other gophers, and avoid a coyote's attack. All these events represent the gopher's interactions with its living and nonliving environment. Ecologists study these interactions at different scales: individuals, populations, communities, ecosystems, and biomes.

POPULATIONS

Populations consist of a group of individuals of the same species in the same area. Characteristics such as size, density, geographic distribution, and age distribution can differ among populations. Individuals might live spread out across a habitat, as in the case of marine copepods, or they might be clustered together, as are mushrooms. The number of individuals at, below, or beyond reproductive age might differ as well. In Italy, for example, just 13.9 percent of the human population is under 15 years old, whereas almost half of Malawi's population falls in the same category.

Populations are dynamic: They grow, shrink, or remain stable, depending on birth, death, and migration rates. However, most populations do not expand

Above: African wild dogs attack a herd of zebras. In the ecosystem of an African savanna, zebras are important prey for predators such as the lion and hyena. Zebras, on the other hand, do not hunt. They are avid grazers, relying on wet and dry grass for their food. Top left: Honey fungus is a common fungus that mostly feeds on dead plants.

exponentially; limiting factors, such as scarce water or common predators, prevent uncontrolled population expansion.

COMMUNITIES

A community is a group of populations of different species in the same habitat. Among populations within a community, various relationships exist, one of which is predation. For example, Canadian lynx hunt for snowshoe rabbits; when the rabbits are abundant, the lynx population grows, but if rabbits are scarce, the lynx are in trouble.

Parasitism, another type of relationship between populations, occurs when one organism uses another for food and habitat. For example, some trematode parasites live within frogs, causing debilitating deformities. Social parasitism is a variation on this theme: The European cuckoo lays its eggs in other birds' nests; these host birds unwittingly incubate and raise the cuckoos' young, sometimes at the expense of their own offspring.

Populations also interact by competing for resources such as space, light, food, water, nutrients, and other materials. Competition is sometimes obvious. For example, a strangler fig wraps a death grip around other trees to win territory. Hyenas and lions fight over a fresh gazelle kill. But sometimes, sneakier tactics are at work. Some plants, such as the black walnut and eucalyptus, exude chemicals into the soil to prevent other plants from colonizing nearby.

Some species within communities have a dominant influence on the structure of the community. These are referred to as keystone species, since they are analogous to the keystone in the center of a stone arch; if the keystone is removed, the arch collapses. In the Pacific Ocean, for example, kelp forest communities are rich with life. Within the complex food web, sea otters are at the top, eating mainly sea urchins. Without sea otters to control their population, sea

Certain birds, such as the Diederik cuckoo, are known as brood parasites. These birds build no nests but instead leave their eggs in the nests of other birds, which then rear the young.

urchins would multiply uncontrollably and graze heavily on their favorite food—kelp. Damage to the kelp could jeopardize the rest of the species within the community. Sea otters, therefore, are considered a keystone species.

ECOSYSTEMS AND BIOMES

Communities within their nonliving surroundings are called ecosystems, which cycle nutrients and water between organisms and their environments. Although physical factors constrain what organisms live in an area, the resident organisms influence their environment as well. For example, vegetation along rivers supports the riverbank and keeps water flowing in a particular direction. When a wolf patrols the riverbank, grazing deer and elk do not graze there, allowing the vegetation to thrive and the riverbank to remain stable. But if the wolf is absent for long periods of time, grazing could jeopardize the riverbank and change the course of the river, affecting downstream life as well.

At the broadest scale are large geographic regions characterized by particular climates and dominant communities, known as biomes. Biomes are useful units for the study of large-scale questions, such as the global transfer of pollutants.

There is a complex web of relationships among the organisms that inhabit an ecosystem. A pack of wolves patrolling a riverbank may prevent elk and deer from grazing there, which in turn allows vegetation to thrive, and the riverbank to remain stable.

A World of Biomes

Diverse landscapes cover Earth. Some regions teem with vibrant life, while others appear desolate and harsh. Availability of sunlight and water combines with geography to influence what organisms live where. Similar species tend to live in similar regions, and these broad geographic regions, which are known as biomes, are dominated by the same climates, communities, and soils. The world's major biomes are generally considered to include tropical rain forests, deserts, grasslands, temperate deciduous forests, boreal forests, and tundra.

RAIN FORESTS AND DESERTS

Tropical rain forests lie at the equator, where the sunlight's duration and intensity are most pronounced. Warm, moist air draws water from the oceans and, as the water vapor cools and condenses in the atmosphere, as much as 33 feet of rain falls each year in rain forests such as the Amazon and the Congo. Dense vegetation creates layers of humid forest, from the dark forest floor to the lofty canopy about 210 feet above. Rain forests are home to more than half of all the world's known species.

Just to the north and south of tropical rain forests, hot deserts stretch out across the continents. Desert lands receive fewer than 10 inches of rain annually. The African Sahara and Kalahari, the North American Sonora and Mojave, the Middle Eastern Arabian, and the Australian Great Sandy deserts all have plants and animals that can endure drought and large temperature swings. Desert-dwelling kangaroo rats obtain all the water they need by eating seeds. The prickly pear cactus has a waxy coating to prevent water loss, and shallow roots to absorb water. Despite their barren appearance, deserts are filled with diverse life.

GRASS AND TREES

Grass is the dominant feature of grasslands, which include subcategories such as steppes, prairies, and savannas. These biomes are found in the slightly wet middle latitudes on every continent except Antarctica. Only slightly more rain falls in grasslands than in deserts—between 10 and 30 inches annually. Trees are sparse because grazing mammals, which eat tree saplings, and periodic fires prevent trees from taking root. The seemingly homogeneous grasses, ranging in height from two to five feet, mask great diversity. Bobcats, eagles, and goldenrod inhabit North American prairies, which are now largely used for agriculture. Steppes in Asia accommodate black-tailed gazelles, bridlegrass, and Henderson's ground jay.

Temperate deciduous forests are found slightly farther north. Well known in the eastern United States, this biome extends across the globe and is characterized by distinct seasonal cycles and markedly different summer and winter temperatures. The resident trees, including maples, oaks, elms, and birch, lose their leaves

Left: The wolverine is found in boreal forests. These northern latitude biomes cover vast stretches of Russia and Canada. Top left: Dense vegetation is a characteristic feature of a tropical rain forest.

in the winter. Some produce seeds that animals disperse. Raccoons, black bears, turkeys, frogs, and many other animals live among the trees.

Boreal forests, or taiga, biomes are found at even higher latitudes, mainly throughout Russia and Canada. Evergreen coniferous trees such as firs, spruces, and pines abound. Cool temperatures keep the region moist and even boggy in some areas, with acidic peat moss. Grosbeaks, loons, moose, and wolverines are some of the animals found in boreal forests.

Arctic and alpine tundra dominate the landscape at the highest latitudes and altitudes. In these biomes, summers are brief and winters are cold. Arctic soil, called permafrost, is virtually permanently frozen. Just a thin top layer thaws to allow plant growth in the summer. Vegetation such as cotton grass is short because it has to grow and reproduce quickly. Snowy owls, tundra swans, caribou, and polar bears make their homes in the tundra.

Polar bears are one of the few species that make their homes in the tundra. Other animals that are able to thrive in this biome include snowy owls, tundra swans, and caribou.

The ground in alpine and artic tundra biomes is frozen most of the year. Only the top layer thaws to allow minimal plant growth.

UNDER THE SEA

Despite this diversity of land, Earth is not really a terrestrial planet. Almost three-quarters of the globe is covered with water, which creates a host of habitats for everything from microscopic dinoflagellates to giant squid.

Saltwater biomes include oceans and estuaries. Estuaries represent the zone where freshwater and salt water blend, and often serve as nurseries for marine species that hide among tree roots or in sheltered coves. Light and temperature diminish at the ocean depths, and life is stratified accordingly. Ocean species such as sharks, tuna, and whales tend to inhabit the upper 1,000 meters of the ocean, whereas the bottom depths are the domain of creatures such as worms and sightless crabs. Even hydrothermal vents at the ocean floor that spew superheated water are the center for a complex ecology.

Coral on the ocean floor. Low temperatures and lack of light restrict the plant and animal life that can survive in the ocean depths. Worms and sightless crabs thrive in this environment.

Freshwater biomes include lakes, rivers, and wetlands. Only 1 percent of the world's water is fresh, flowing water, but the survival of trout, cattails, green algae, crustaceans, dragonfly larvae, and countless other living things, including humans, depends on this source.

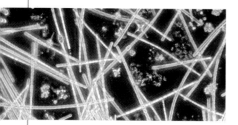

Who Eats Whom?

Figuring out predator-prey relationships—who eats whom—is a meaningful way to describe relationships among living things in an ecosystem. Diagrams called food chains or food webs illustrate these relationships. A food chain is a simplified version of feeding relationships. For example, grass seeds are eaten by mice, which are in turn eaten by hawks.

In reality, feeding relationships are much more complex and better illustrated by a food web, which shows the various organisms that can eat or be eaten by many others. A mouse, after all, is not dinner for just a hawk; a fox, falcon, or any number of other predators can dine on a mouse morsel. Similarly, feeding relationships can change throughout an organism's lifetime. A tadpole and a frog occupy different places in the food web.

LEVELS OF THE FOOD WEB

At the bottom level of all food webs are photosynthetic producers, such as plants and algae, which convert sunlight energy into food energy. Next are plant eaters (herbivores), such as grasshoppers, squirrels, and deer. These are followed by first-, second-, third-, fourth-, and fifth-level consumers, including coyotes, snakes, crows, and hawks.

Corresponding levels are found in lakes and oceans, with phytoplankton as producers, grazers such as copepods as first-level consumers, and shrimp, bluefish, tuna, and sharks as higher-level consumers. Some food webs cross the terrestrial-aquatic boundary: Seagulls dive for fish, for example, and some fish leap out of the water to catch airborne insects. The food chains teased apart from food webs generally have five or fewer levels. Also, the number of organisms at each level decreases up the food chain.

TRANSFERRING ENERGY AND POLLUTANTS

Although food webs most obviously show who eats whom, they also explain the transfer of energy through an ecosystem. The Sun provides the Earth with a huge amount of energy, but only a small fraction of that energy is captured and used by plants. A fraction of the energy stored in plants is eaten and

Above: A mouse, which is just a step removed from the bottom level of the food web, dines on a seed. Top left: Phytoplankton are the bottom level producers in lakes and oceans. Left: A hawk fulfills its role within the food chain when it feeds on a mouse. It is, however, one of several animals that prey on mice. The number of organisms at each level decreases up the food chain.

Flies crawl over a pile of barren corn cobs. Flies are considered decomposers because they help to break down wastes and dead organic materials.

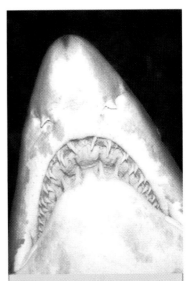

The razor-sharp teeth of a great white shark make it one of the fiercest predators of any animal species.

used by first-level consumers, and so on, but energy is lost at each step. The plants in an ecosystem represent more energy than first-level consumers, which represent more energy than the highest-level predators. Similarly, the number of individuals at each level is greater than the number at the level above it.

Unlike energy, some compounds are not lost at each step—and they accumulate as they travel up the food chain. Chemical pollutants such as DDT and mercury fall into this category and become more concentrated in higher-level consumers. Because tuna are apex, or top-level, predators—animals that are hunted by few if any other natural predators—their mercury levels can be particularly high.

However, sardines from the same waters would probably have much lower contamination. You are what you eat, for better or worse.

DECOMPOSITION

Where do decomposers fit into the picture? Bacteria and fungus break down wastes and dead organic material. In a food web, they are often shown as the link between the top-level predators and the producers because they recycle nutrients.

Alternatively, decomposers can be considered part of the food web as consumers. Sometimes, they are not included in food webs at all; this is an oversight, because no matter how they are classified, they play a crucial role in making nutrients available for other organisms.

BORN PREDATORS

Predators at the top of the food web are known for jaws, claws, speedy attacks, and cunning behavior. Some are solitary hunters, while others hunt in groups. From the point of view of their prey, however, all are fearsome.

Peregrine falcons sport sharp talons for tearing apart rodents and smaller birds and can dive through the air at nearly 70 miles per hour when pursuing prey. Mountain lions stalk deer, elk, and other animals, bursting into a 40-mile-per-hour pursuit and breaking a victim's neck with a single bite equivalent to 900 pounds of force.

Great white sharks are eating machines that come equipped with about 50 mature teeth and a replacement reservoir of as many as 500 developing teeth. At an attack speed of 25 miles per hour, a shark is hard to outswim.

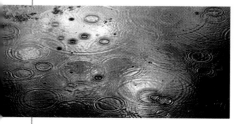

Unending Cycles

In nature, nothing is wasted. Everything is recycled. Materials such as oxygen and potassium are moved from the environment into organisms and back to the environment. In diagrams, these cycles are often shown as simple circles, but they are in fact quite complex, with materials entering and exiting at multiple points. Water, carbon, and nitrogen cycles are very influential to living things. These three materials, all of which are essential for life, constantly circulate in and out of the bodies of living things.

WATER CYCLE

The water cycle begins with evaporation from the oceans and land plants. The vapor ultimately returns to the Earth's surface as rain or snow. Precipitation that falls on land seeps through the ground or travels by gravity over the surface by a watercourse such as a river, stream, or creek. The water from a particular area ultimately drains into one river or stream. This watershed can be as small as a few acres or as large as a third of the United States, as in the Mississippi River watershed.

The water cycle is vital because water is required for life and because it transports minerals such as calcium, which is essential for muscle contraction, nerve transmission, blood clotting, and growth of bones and teeth. For human consumption, the water cycle must be monitored to ensure adequate supplies despite seasonal changes of inflow and outflow.

CARBON CYCLE

Like the water cycle, the carbon cycle is also crucial for life. Carbon is the principal building block for all living things. It is found in DNA, proteins, carbohydrates, and fats. Carbon enters the atmosphere, mostly

Right: The Earth's total water supply is constantly redistributed by a variety of processes. Water evaporates from the ocean and may fall as precipitation over land (lower left, top). Water reenters the atmosphere from land by transpiration (water loss by plants) and evaporation (top). Water flows from land to ocean in surface rivers and streams (center). It also travels through soil and rock as groundwater (right). There is 30 times as much fresh water below ground than in lakes, rivers, and inland seas. Top left: Rain forms a pool of water.

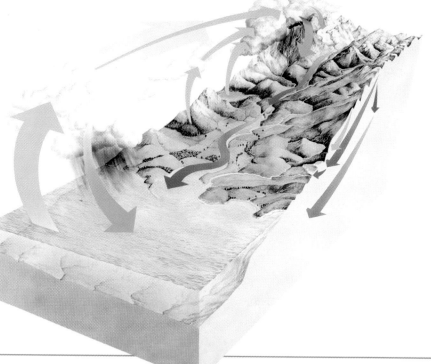

A tractor trailer spews smoke from its exhaust pipes. The smoke is a carbon byproduct in the burning of fossil fuels. Many scientists believe that the burning of fossil fuels is contributing to an abnormal warming of the Earth.

as carbon dioxide, from cellular respiration, volcanic eruptions, and the burning of fossil fuels and wood. In oceans, it is mostly dissolved as bicarbonate and carbonate. Photosynthetic plants and algae absorb the carbon and incorporate it into their molecules. As plants are eaten, carbon moves through the food web and is transferred to other organisms. As plants and animals die, their decomposition liberates carbon for recycling.

Algae blooms in a stagnant pond. Algae plays a key role in the carbon cycle, using the energy of light to convert carbon dioxide into organic matter.

The burning of fossil fuels and the destruction of forests (deforestation) add carbon to the atmosphere much faster than it is used by living things, which creates a buildup of atmospheric carbon in the form of carbon dioxide. There it functions as a greenhouse gas, trapping heat from the Sun on the planet's surface. The general consensus from atmospheric scientists is that this process is contributing to the abnormal warming of the planet (global warming), as compared to what would be expected without the influence of human activity.

NITROGEN CYCLE

Nitrogen, a component of the amino acids that make up proteins, is another element required for life. Nitrogen

constitutes 80 percent of the atmosphere, making it seemingly an abundant resource. In fact, it is often the scarcest nutrient in ecosystems. Nitrogen must be fixed into molecules that plants and animals can use, which only lightning, volcanic activity, and certain bacteria can do. Plants recruit nitrogen-fixing bacteria to live near their roots by providing them with sugars. Once bacteria integrate nitrogen into ammonia and ammonium, it is available for other organisms to use. When plants and animals die, bacteria break up molecules and release gaseous nitrogen back into the atmosphere.

Nitrogen is a major component of fertilizer, and more than 50 million tons of fertilizer are used in the United States every year. Although the nitrogen used in fertilizer is cycled through plants and eventually released as gaseous nitrogen, deforestation removes nitrogen from the ecosystem completely.

A Role for Every Player

Each species has a different role to play in its ecosystem, and that role is called a niche. A niche is a way of life, a way in which a species can use its particular set of characteristics and abilities to survive and reproduce in its environment. Those characteristics and abilities are its adaptations. As ecologist Eugene Odum put it, a niche is analogous to a profession.

ADAPTATIONS

Each species adapts to live within its physical and chemical environment. Penguins have thick layers of blubber to keep them warm in the cold-water niche where they fish. Giant anteaters exploit a niche by using their two-foot-long tongues to lick up ants from their nests.

Sometimes it is not immediately apparent why an organism occupies its particular niche. For example, two species of barnacles can live in the same intertidal zone, between the high-tide and low-tide marks of a shoreline. But one is always found on higher, drier rocks, whereas the other lives on the lower, wetter rocks. But the offspring (larvae) of both species float through the water and settle everywhere. To discover the reasons for these particular habitat selections, ecologists removed some of the barnacles that lived on the lower, wetter rocks; they observed that the barnacles that favored the higher, drier rocks migrated to occupy the newly created open

space. But when the ecologists removed some of the barnacles from the higher rocks, the low-rock species did not migrate, as they could not survive on the drier rocks.

This experiment showed that the upper-rock species could live all over the rocks, if space was available. Under ideal circumstances, it would live anywhere on the rocks, a way of life known as its fundamental niche. But with the other species present, it was restricted to the upper rocks. This is the barnacle's realized niche. This phenomenon is sometimes known as competitive exclusion.

SUCCESSION

Communities change over time in a process called succession, which is related to the development of niches. When a disturbance—a volcanic eruption or a glacial retreat, for example—creates space for new life, new communities develop in a somewhat predictable sequence. When rock is virtually bare, the first colonizers are lichens and mosses. These inhabitants help accumulate soil. When soil is present, plants can move in and crowd out the original settlers. This process can be repeated several times over. Succession can also occur on rooftops of neglected houses and in abandoned

Above: The penguin's thick layer of blubber is an adaptation to its frigid environment. Top left: Barnacles on a rock. Ecologists have discovered that not all species of barnacles can survive on both dry and wet surfaces.

fields—over many years, each of these environments could again harbor its original community.

As glaciers retreat in Alaska, lichens soon cover bare rocks, followed by horsetail and fireweed. Over time, alder, cottonwood, and willow move in; they stay for decades, until mature thickets are speckled with spruce. A century later, spruce and hemlock reign. This is succession in action.

REASONS FOR SUCCESSION

Succession happens because resource availability changes as a community changes. When lichens take hold, rock is no longer simply rock. Soil collects and becomes a resource for seeds. As taller trees develop, they shade the understory below. Life changes the physical properties of the environment. Each stage of succession opens up a new set of niches.

Although succession is often depicted as a linear process, it does not always proceed in a linear fashion. Communities are frequently disturbed to varying intensity by storms, rockslides, floods, and other events. Succession can take many paths. When disturbances eliminate some populations or species, the remaining survivors might return or colonizers from adjacent regions could migrate in.

The aftermath of a destructive forest fire is a prime opportunity for different species to colonize the cleared area. Some species of trees return readily because their root systems remain intact underground.

A FIERY ORIGIN

Fires are an immensely destructive force. Wildfires often ignite when a lightning strike hits dry pine needles, leaves, branches, and dead snags. The blaze can whip through a vast area or burn more slowly, consuming everything in its path.

But out of this fiery destruction comes ecosystem rejuvenation. By creating a clearing, fires "reboot" succession, allowing different species to colonize the area. In the American West, sagebrush, aspen, and willows return after a fire because their root systems stay intact underground.

Some species are adapted to fire. Thick bark protects Douglas firs from intense heat. Some lodgepole pines have cones sealed with resin that melts during fires. Seedlings that sprout from the newly opened cones thrive in sun-drenched clearings. These same seedlings would not have a chance of success in a mature, dark forest.

Top: The purplish pink flowers of the rosebay willow herb, also known as fireweed, are a common sight in cleared woodland and wasteland. Bottom: Moss and mushrooms sprout from a branch. These plants are among the first successors in new communities.

Partners in Survival

Many species interactions benefit one species at the expense of another. Predation and competition are two examples. But nature also affords elegant examples of cooperation between species. Living things that are intertwined in mutually beneficial relationships are called mutualists. Relationships between mutualists can be necessary to the survival of one or both species (an obligate relationship) or optional.

Mutualism is one kind of symbiosis, an interaction in which members of two different species live together in a close relationship for the benefit of at least one member. Both members may benefit, as in mutualism. But one member can be harmed, as in parasitism. A parasite benefits from another organism (its host) while harming the host. An example is the hookworm, which lives in the intestines of human beings and other animals and sucks their blood, making them ill. In a third kind of symbiosis, called commensalism, one member benefits while the other remains unaffected by the relationship.

MUTUALISM

Yucca plants and yucca moths have a mutualistic relationship. Each species of yucca moth must lay its eggs in a particular species of yucca plant. The flower of the plant provides protection for the eggs, and, once the larvae emerge, nourishment with a few extra seeds. In exchange, the moth pollinates the flower.

Other examples of mutualism abound in nature. African wading birds remove parasites from crocodiles' teeth and enjoy a meal in the meantime. Acacia trees woo ants by oozing sugary sap; the ants in turn protect the trees from other insect attacks. Termites are able to digest only wood thanks to protists living in their guts. Lichens are not actually one organism but a composite of mutualistic fungus and green

Above: Two yucca moths pollinate a yucca plant's flower. The plant and the moth benefit from each other's existence, with the yucca plant providing vital protection for the moth's eggs. There are many instances in nature where the relationship between two species is necessary for both of their survival. Top left: The acacia tree maintains a mutualistic relationship with ants. Its sugary sap attracts ants. In turn, the ants protect the tree from other insect attacks.

A clownfish is protected from predators by a stinging anemone. This is an instance of commensalism, in which only one party benefits from a symbiotic relationship. It is not clear what benefit the anemone derives from the fish.

algae. The algae acquire a habitat in exchange for providing food to the fungus.

COMMENSALISM

In commensalism, only one party in the symbiotic relationship benefits; the effect on the other is neutral. Such is the case with egrets in Africa, which feed near grazing cattle. As cattle feed, they kick up insects, which the egrets eat. The egrets are neither a help nor a hindrance to the cattle.

It is not always clear whether both parties in a symbiosis benefit or whether only one does. On coral reefs, for example, clownfish live within the protection of stinging anemones. There is no definite advantage to the anemones in the relationship, but it is possible that clownfish lure prey for the anemones.

HUMAN MUTUALISM

Humans have mutualistic relationships with many bacterial species. We provide habitat and nutrients for them in exchange for necessary services. Helpful bacteria in our mouths and noses take up space, preventing disease-causing species from colonizing those areas. In the large intestine, bacteria break down carbohydrates, proteins, and fats that were not digested earlier in the gastrointestinal tract. These smaller components can then be absorbed by the body.

In women, useful bacteria maintain a vaginal pH level of about 4.5, which thwarts yeast infections. Sometimes, as an unintended consequence of modern medicine, antibiotics taken for illnesses kill helpful bacteria, creating opportunities for infections.

Not all symbiotic relationships are mutually beneficial as is the case with parasitism. The hookworm parasite, for instance, invades the intestines of animals, making them ill.

The Human Footprint

Above: Pesticides, fertilizers, and agricultural waste can leach easily into the groundwater and pollute drinking water. Top left: Sulfur dioxide emissions from factories accumulate in the atmosphere and are a source of acid rain.

What mark are humans leaving on the Earth? How much are we changing its ecology? The ecological footprint is a concept that describes our impact on Earth. One way of viewing an ecological footprint is as the area of land and amount of water needed to support a human population at a particular standard of living.

Every living thing manipulates its environment. Beavers build dams that redirect water sources. Elephants tear down trees. Seabirds create islands of guano. People, too, manipulate the environment, but at a magnitude and frequency unparalleled by wild species.

PHYSICAL AND CHEMICAL CHANGES

Humans alter their environment in many physical, chemical, and biological ways. Physically, we

Overgrazing and overfarming have led to serious consequences for the environment, converting fertile land into desert.

cut down forests, which destroys habitats and removes support from the soil, causing erosion. In addition, deforestation removes a nitrogen source from the soil. We divide habitats by building highways and developments, preventing mammals, birds, amphibians, and insects from migrating along their customary routes. Diverting and damming rivers prevent fish, such as Pacific salmon and white sturgeon, from spawning. Overuse of water resources has lowered water tables. Overgrazing and overfarming can convert formerly fertile land into desert, as illustrated by the 1930s Oklahoma dust bowl and the current desertification in the West African Sahel.

Chemically, pesticides, fertilizers, and agricultural waste leach into the groundwater. Factories emit sulfur dioxide into the atmosphere, which creates

acid rain. In New York State alone, acid rain has led to the disappearance of fish from more than 200 lakes. The burning of fossil fuels releases carbon dioxide into the air, contributing to global warming.

BIOLOGICAL CHANGES

Human activity has been causing species extinctions at an accelerating rate. In the 1800s, North American passenger pigeons were among the most common birds on the continent, numbering in the hundreds of millions. By 1914, they were extinct due to overhunting. In the late 1800s, whalers killed the last Steller's sea cow. The Pyrenean ibex and the West African red colobus monkey were declared extinct in 2000.

Humans also move species around the globe, intentionally and accidentally. Some, like the zebra mussel, hitch a ride as ballast on an ocean freighter and then significantly alter their new habitat. Others, like the kudzu that grows uncontrollably in the American South, were introduced by well-meaning horticulturists. Invasive species (those that thrive and disrupt their new communities) are second only to habitat destruction in causing species extinctions.

EFFECTS OF THE HUMAN FOOTPRINT

Taken independently, these ecological changes might seem trivial. But each extinction, species introduction, or physical change causes a perturbation in an ecosystem, with ripple effects far beyond the immediate surroundings. No one would have guessed that a human-induced decline in England's rabbit population would set off a chain of events that ended in the extinction of a butterfly. But grass that the rabbits had eaten flourished in their absence, reducing ant populations that required open land to build their mounds. Without the protection of the ant tunnels, butterfly larvae that would have developed there perished.

The services that ecosystems provide to human beings are sometimes overlooked, difficult to quantify, or undervalued. But those services are many. Ecosystems absorb carbon dioxide, emit oxygen, filter water, convert solar energy into food energy, fix nitrogen, regulate stream flows, buffer coastlines from storms, offer a potential source of pharmaceuticals, provide recreational opportunities, and create tourism jobs. The loss of a single small species might not influence the tourist-based Kenyan economy, for example, but the loss of the big mammals—rhinoceroses, elephants, and leopards—would have dire consequences.

With the human population well beyond six billion, more people around the globe are becoming aware of the need to reduce the human footprint on the world's ecosystems, or at least manage it more wisely. The study of ecology provides the knowledge needed for that task.

Unchecked logging destroys habitats and leads to erosion. The resulting loss of trees also removes a crucial source of nitrogen from the soil.

Dodos became extinct in about 1680.

FAREWELL TO THE DODO

The dodo was named by sailors who arrived on the island of Mauritius in the Indian Ocean in the 1600s and were dumbfounded by the awkward bird. It was about the size of a large turkey, with a hooked beak and stubby wings. It was clumsy, could not fly, and seemed irrationally fearless. But it had cause to fear.

The men killed some birds, possibly for eating. The nests of those remaining were then decimated by pigs and other animals introduced to the island by the sailors. The dodo became extinct in about 1680, a tragic example that highlights the crushing power of the human footprint.

Centuries later, an island tree, *Calvaria major*, became rare, and scientists hypothesized that the disappearance of the dodo was to blame. Without the dodos to swallow the fruit and excrete the seeds, the seeds could not germinate. Today, turkeys or gemstone polishers are substituted to do the job of processing the seeds.

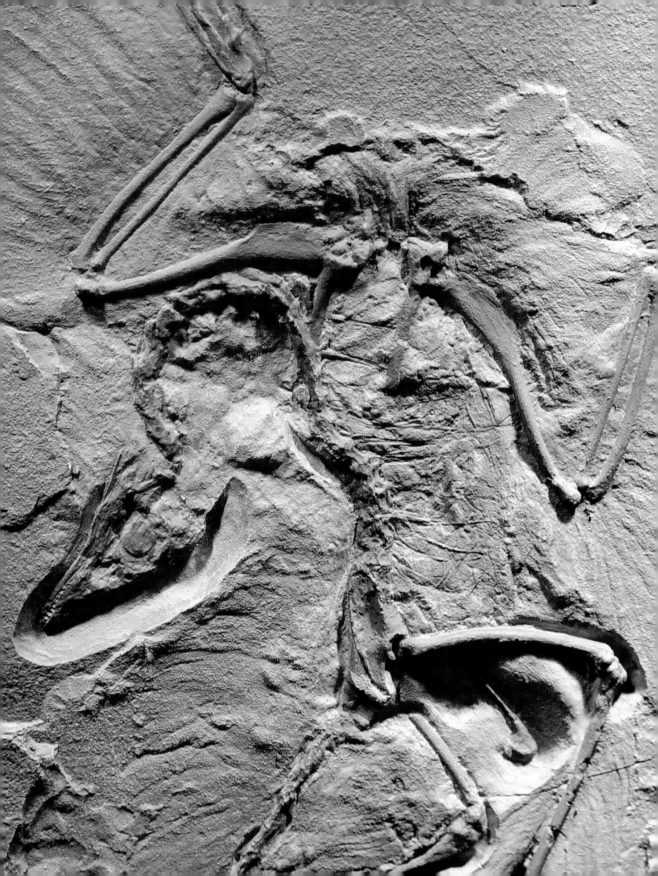

CHAPTER 8

THE EVOLVING TAPESTRY

Left: This Archaeopteryx fossil dates from the late Jurassic period about 150 million years ago. Fossilized remains of ancient organisms allow scientists to create a record of the prehistoric past. Due to the reptile features of this ancient bird, scientists believe the Archaeopteryx is an important link to understanding how dinosaurs evolved into birds. Above: Yellow roses. Genetic mutation has caused the rose bush to produce flowers in varying shades of yellow. Bottom: A kangaroo with its offspring. With the decline of dinosaurs, mammals evolved into larger and smarter creatures. They filled niches in the animal kingdom once occupied by dinosaurs.

Life has a history. The landscape of living things has not always looked the way it does today, harboring the vast number of species we know. Once, there were no animals that could breathe air or plants that could flower. The ancestors of birds were unable to fly. There were giant reptiles as big as houses, humanlike creatures who stood erect but were not *Homo sapiens*, and shelled animals with eye lenses made of limestonelike calcite.

The study of past forms of life such as these is called paleontology, and the study of how one form of life gives rise to another is called evolutionary biology. Evolution is the engine of biological change, the force that writes the history of life. Evolution is descent with modification: the generation of a line of offspring in which some characteristics of descendants are not exactly the same as those of their ancestors. When characteristics change enough, entirely new species take shape.

Evolution relies on the genes that organisms inherit from their parents. It processes those genes through mechanisms that include natural selection and genetic drift. The result of evolution is the history of life as spelled out in the fossil record, from Precambrian organisms through the first humans.

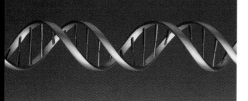

Inheritance— Material of Evolution

The basis of biological evolution is the gene. A gene is a unit of heredity, an instruction transmitted from parent to child that shapes the characteristics of the child. These units of biological inheritance are composed of deoxyribonucleic acid, or DNA. In eukaryotes such as plants and animals, genes are organized into chromosomes, which reside in the cell nucleus.

If genes transmitted parental information perfectly, there would be no evolution. After three billion years, nothing would exist except more of whatever DNA-carrying organisms existed at the beginning. But genes are not perfect transmitters of information. They get copied incorrectly and get shuffled in new combinations. Over three billion years, the effect of these imperfections on Earth has been the diversification of bacteria into fish, trees, ants, elephants, human beings, and the myriad other inhabitants of the planet. The imperfect inheritance of characteristics through genes is the material of evolution.

GENES ON THE MOLECULAR LEVEL

A gene is a segment of a DNA molecule in a chromosome. It gets its peculiar ability to transmit inherited information from the structure of the DNA molecule. The DNA molecule is a double helix, a shape similar to a twisted rope ladder. Each rung of the ladder consists of two matching compounds called bases. There are four bases: adenine, guanine, thymine, and cytosine. These bases match up in a specific way:

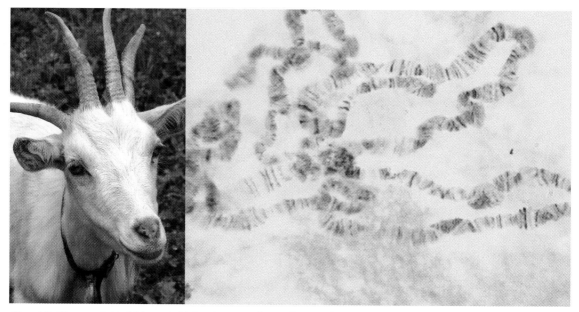

Above left: The cause behind this goat with four horns can be traced to genetic mutation. The gene, a unit of heredity, is the driving force of biological evolution. Above right: All plants and animals have chromosomes that each consist of a single molecule of DNA. Top left: A rendering of a DNA double helix molecule. The molecule, which has the appearance of a twisted ladder, gives chromosomes the ability to transmit hereditary data.

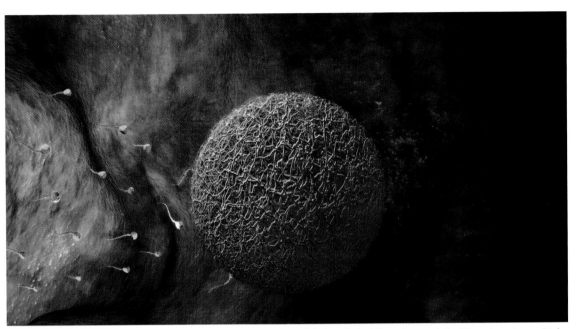

Illustration of a sperm cell fertilizing an egg. The laws of genetics help explain how the hereditary characteristics of parents are transmitted to offspring. The process is not random. Rather, the transmission of genes during sexual reproduction follows certain patterns. The resulting shuffling of genes following their transmission, however, produces a unique individual. This shuffling follows an unpredictable path.

adenine will pair up only with thymine, and guanine will fit only with cytosine.

During cell division, the DNA molecule splits in half, dividing the base pairs. Each separated strand of the DNA molecule then pairs with the corresponding complementary bases to complete the double helix. A cytosine base will attract only a guanine base, just like the one to which it used to be joined. The result is the building of two complete copies of the DNA molecule where before there was only one.

Not only can DNA reproduce itself; it can direct the creation of proteins, the building blocks of organisms. This is possible because a particular string of DNA segments, called a codon, corresponds to only one of the 20 amino acids present in body proteins. A long chain of DNA can therefore represent a coded set of instructions for a sequence of amino acids that makes up a particular protein. It can also transfer those coded instructions to the parts of the cell that can assemble that protein.

In this way, the DNA of an offspring reproduces the genes of its parents and directs the development of the offspring based on those genes. This all happens with a rate of error that is amazingly low, yet higher than zero. Extended over eons, this low error rate drives evolution.

LAWS OF GENETICS

When transmitting inheritance, genes follow patterns known as the laws of genetics. The first law, called the law of segregation, states that each hereditary characteristic is controlled by two genes, but that the genes in a pair segregate (separate) when sex cells are formed, so that each sperm or egg receives only one member of the pair. According to the second law, the law of independent assortment, each pair of genes breaks up independently of the other pairs when sex cells are formed, so that the exact assortment of genes in any given sex cell cannot be predetermined. During sexual reproduction, the genes contributed by the male and the genes contributed by the female form a new set of genes.

The shuffling that results from the laws of genetics is so thorough that each individual is born unique, carrying the genes of its parents in an unpredictable new combination.

The Evolution Toolkit

Evolution uses different tools to modify organisms. One tool is genetic variation, a condition that is essential to the working of the other tools. Another is natural selection; first identified by Charles Darwin, this mechanism adapts populations to their environment. Finally, there is genetic drift, random changes in the frequency of genes that do not tend to improve adaptation.

GENETIC VARIATION

Without genetic variation, evolution would be impossible. Genetic variation is driven primarily by three occurrences. One is mutation: a change in an organism's DNA. Such changes happen infrequently but regularly when a dividing cell fails to copy its DNA correctly. The presence of radiation or certain chemicals can spur mutation to happen more often. If the changes take place in sperm or egg cells, where they can be inherited by the next generation, the effect of even small mutations can accumulate over time to become significant.

Another engine of gene variation is gene flow, the movement of genes from one population to another. Also known as migration, it includes pollen moving from one location to another and the movement of people from one place to another. Even in populations in which there is no migration, sex is a third source of genetic variation. The new DNA combinations that arise in each act of fertilization produce new gene combinations.

Left: The color of the white owl's feathers allow it to elude predators and succeed in capturing prey. The natural selection process has given the white owl the tools to survive and reproduce. Top left: A river serves as a conduit for tree pollen.

NATURAL SELECTION

Genetic variation by itself does not tend to improve the adaptation of living things to their environment. But natural selection does. Natural selection depends on the power of genes to determine the observable traits (or phenotype) of organisms—for example, whether they are large or small, or dark or white.

Because genes vary in a population, observable traits also vary. However, not all traits are equally advantageous. Scarcity of food, the stalking of predators, and other factors tend to weed out the individuals whose traits are less beneficial. By contrast, individuals with traits well suited to the environment—that is, individuals that are more fit—survive and reproduce. For example, if dark hares are more visible than white hares against a snowy landscape, owls and other predators will tend to eat the dark hares and miss the white ones.

The winners of this struggle for survival (the white hares) pass on their genes for greater fitness to their offspring, while the genes of those that die before reproducing (the dark hares) are eliminated. In this way, the more adaptive traits tend to become more common in the population. Over time, refinements that make the traits even more advantageous are added through the same process, and whole new species form. The snowshoe hare, which has the ability to grow white fur in winter to hide itself from predators, is an example of an animal that evolved through this process.

A special case of natural selection is sexual selection. In this process, organisms that have traits that attract the opposite sex are more successful in finding mates and passing on their genes to offspring. As a result, genes for those attractive traits—such as the peacock's tail—become more common.

GENETIC DRIFT

In genetic drift, as in natural selection, genes become more or less common from one generation to the next. Unlike natural selection, changes in genes' frequency from genetic drift do not make a population more or less adapted to its environment. Some individuals may leave a few more descendants than others—by chance. These individual's genes will then become more common. Island populations may change through genetic drift because they began with a different or smaller subset of variation than mainland counterparts.

Lava flow is a result of tectonic activity, a key mechanism in the evolution of species.

PLATE TECTONICS AND EVOLUTION

One way a new species evolves is if a group of individuals gets cut off geographically from other individuals of the same species. The separated groups may evolve into entirely different species. A principal mechanism of this process is tectonic activity.

Earth's outer shell is composed of tectonic, or structural, plates. These large sections of rock, which have the continents embedded in their tops, move around constantly on the asthenosphere, a hot layer of rock below. Over the course of ages, the drifting of plates breaks and assembles continents, raises mountain ranges and volcanic islands, and opens rifts.

Tectonic activity is a perpetual spur to evolution because of the way it isolates populations. After breaking off from South America, Australia developed kangaroos, koalas, and other unique marsupials. When volcanic activity in the Pacific Ocean formed the Galapagos Islands, it spawned a habitat for birds, iguanas, and whatever other animals could reach it from the mainland. These animals evolved into species found nowhere else on Earth.

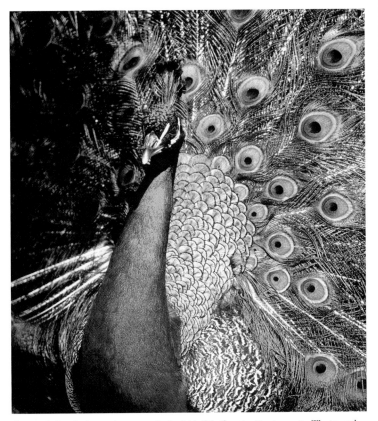

A peacock fans its impressive array of colorful tail feathers to attract a mate. The peacock gained this trait through sexual selection, a special instance of natural selection.

Evolution— The Evidence

Evidence that living things evolve comes from many sources. Some of these sources are no longer alive: namely, the fossilized remains of organisms from the prehistoric past. Some evidence of evolution is found in living bodies themselves, for example, in the resemblances between the DNA of different species. Various experiments in evolution, natural or human-made, further testify to the power of evolutionary mechanisms.

FOSSILS

It is not possible to look directly at the plants and animals that lived long ago, but it is possible to examine their remains or traces when they are preserved in rock (fossilized). The age of a fossil can be determined through radiometric dating, which is based on measuring how much the radioactive atoms inside a rock have decayed since the time the rock was formed. Since radioactive decay happens at pre-dictable rates, this measurement reveals how old the rock is.

The many fossils that sci-entists have discovered and dated offer an incomplete but informative record of what types of organisms lived and when they lived. This record supports the view that living things evolve. The fossils reveal a past in which

there were once only simple forms such as bacteria, followed by the appearance of progres-sively more diverse forms of life. Much of the early life is now extinct, and many early species resemble later ones as if they were their ancestors.

THE EVIDENCE OF LIVING BODIES

More evidence for evolution is found in the bodies of living or-ganisms, encoded in their DNA. Random changes in a sequence of DNA generally happen at a regular rate over time, establishing a molecular clock. By measuring differences in a shared DNA sequence in two organisms (such as a human and a chimpanzee), scientists can infer that a certain amount of time has passed since their ancestors began to evolve into different species.

Embryology (the study of how organisms develop in their early stages) also supports the theory of evolution. Embryos pass through stages in which they resemble other species, as if the animal has retained some of the developmental features of an ancestor. For example, a human fetus has a coat of fine hair called the lanugo, which it sheds before or soon after birth. Ape fetuses grow a similar coat of hair but do not shed it.

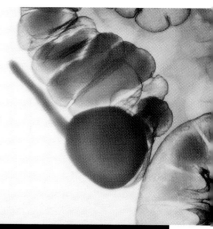

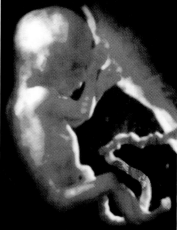

Top: The narrow, finger-shaped human appendix has no known function. Bottom: A human fetus is covered with a coat of hair, which is shed soon after birth. Ape fetuses have a similar coat but do not shed it. Top left: A fossil is carefully excavated.

Some holdovers are not temporary. Living things have organs that are useless or even harmful to them, such as the tube

A WHALE WITH LEGS

The most telling fossils in support of evolution are those of transitional forms, which are organisms that combine the characteristics of two types of organisms. These individuals are intermediate forms between one type of organism and another. For example, at one point in the fossil record, no fossils of mammals have yet been found, while at a later time, reptile fossils that have some resemblance to mammals are evident. Moving forward in time, the resemblance between reptiles and mammals is more pronounced. Finally, still closer to the present day, fossils of mammals are apparent.

Another example of a transition involves the whale. Scientists have long believed that whales descended from land mammals, but for a long time there was little solid evidence of forms that were intermediate between land mammals and whales. Then, in Pakistan in 1994, paleontologists discovered the remains of a 50-million-year-old whale, *Ambulocetus natans*. It had large hind legs that were functional on land and in water—the long-sought-after transitional form.

Ambulocetus natans, ancestor of whales, lived about 50 million years ago. Its hind legs were functional on land and in water.

attached to the large intestine that is the human appendix. The existence of such organs makes sense when they are viewed as vestigial organs, useless relics of structures that were useful in our ancestors. They still are useful in orangutans and other apes, for example, in which the appendix is a functional sac that helps digest plant material.

EXPERIMENTS IN EVOLUTION

Direct observation of evolution through experiment is difficult because evolution happens slowly, over many generations. Nevertheless, some natural and human-made experiments provide evidence for evolution.

Biogeography (the geographic distribution of species) is a kind of evidence based on natural experiments that took place in the past. For example, certain island groups, such as Hawaii and the Galapagos Islands, arose from ocean volcanoes and have never been connected to continents. Their native species are restricted to those that can travel easily over large stretches of water, such as bats, birds, and small animals that might have floated there on driftwood. Yet those species often take unique forms, suggesting that they have evolved since their arrival.

Scientists have begun to duplicate nature's experiments in evolution with some of their own. In one study, lizards were placed on 14 different islands that were normally devoid of lizards. In as little as a decade, the island populations had begun to evolve different adaptations, such as longer or shorter legs.

Marine iguanas have evolved since they arrived on the Galapagos Islands.

Mass Extinctions

Evolution is always occurring, but not always in the same way. It often happens at a slow and steady pace, as genes drift in frequency and small improvements to fitness are naturally selected. But it can also happen in a more rapid, sudden burst. One way that this occurs is through a mass extinction.

A mass extinction is the elimination of many species in a short time. "Short" is a relative term: In geologic time, this period may be several million years. But a mass extinction is sudden compared to the usual, stately pace by which species appear and disappear. For the species that are eliminated, mass extinctions are a disaster.

But for the species that survive, they are golden opportunities for evolution.

CAUSES OF MASS EXTINCTIONS

Most of the mass extinctions in Earth's history have had natural causes. Climate change was often the culprit. Ice ages, periods when glaciers covered large regions of the globe, have suddenly made many environments unusually cold. The effect has been to kill off any species that could not tolerate the change. Such an ice age happened at the end of the Ordovician period, 443 million years ago, causing a mass extinction.

Asteroids or comets have sometimes struck the planet, with catastrophic results (see sidebar). Sometimes, Earth-bound tectonic activity has been the trigger for mass extinction. At the end of the Permian period, 248 million years ago, the largest extinction in Earth's history occurred. It wiped out at least 80 to 90 percent of all species, including about 95 percent of all marine species. An asteroid or comet might have been involved, but there is also evidence that volcanic activity was at work. Eruptions were pouring lava all across Siberia, producing enough debris to block sunlight and chill the climate.

Above: An iceberg in the South Atlantic. Top left: A water-filled crater. Asteroids and ice ages have been responsible for mass extinctions.

An asteroid may have killed off the Tyrannosaurus rex and all other dinosaur species.

Top: The giraffe, a plant-eating animal, has filled the void left by large herbivore dinosaurs. Bottom: Mammals in the dinosaur era were small like the modern-day rat.

DEATH OF THE DINOSAURS

Sixty-five million years ago, at the end of the Cretaceous period, an asteroid estimated to be six miles wide slammed into Earth, digging a crater more than a hundred miles wide off the coast of Mexico's Yucatán Peninsula. The impact released 100 million megatons of energy (the energy released by a nuclear bomb is on the order of 20 megatons), generating tsunamis, earthquakes, hurricane-force winds, acid rain, and possibly a global heat pulse that sparked fires around the world.

But the worst effect was a cloud of dust suspended in the atmosphere. The cloud caused total darkness globally for weeks or months, then twilight conditions for several more months. A shutdown of photosynthesis resulted, making food webs everywhere collapse.

Most scientists now believe that this asteroid impact caused the extinction of the dinosaurs, along with almost all of the planet's large vertebrates and many land plants. It is estimated that about 70 percent of all species became extinct. Mammals survived, perhaps because they were small and omnivorous enough to make do with whatever scraps they could find in the dark times.

In addition, the formation of the supercontinent Pangaea at this time, as a result of tectonic activity, may have changed ocean patterns and wind currents in ways that disrupted ecosystems.

Mass extinctions are not just episodes in the past. Many scientists think we are currently in the midst of a human-made mass extinction that may turn out to be the biggest since the time of the dinosaurs. The causes of this extinction include global warming from carbon dioxide and other waste gases; pollution; poor land management; destruction of wildlife habitats; overharvesting of animals and plants; and introduction of invasive, or non-native, species. As a result, some scientists fear, nearly a quarter of all living species are in danger of becoming extinct within the next 50 years.

ADAPTIVE RADIATION

Although mass extinctions are disastrous for the species that die off, those that survive are left with opportunities. Species with which they once competed are gone, as are predators that once attacked them. Niches that were once filled by other species are now empty. With plenty to eat and little to fear, the survivors swell in numbers. Their numerous offspring venture into niches that used to be occupied by better-adapted species. For example, before the disappearance of the dinosaurs, mammals were small, somewhat akin to rats of the present day. But the extinction of the dinosaurs, until then the dominant life form, left mammals with the chance to evolve to fill all the dinosaurian roles, from large herbivores (elephants and giraffes) to large carnivores (modern lions and tigers).

This kind of explosive increase in the number of species in a group is called an adaptive radiation. It might be seen as the bright side of a mass extinction.

How Life Began

Earth arose 4.6 billion years ago from the clumping together of dust and condensed gas as the planets took shape around the infant Sun. For hundreds of millions of years, Earth remained lifeless. Approximately 4 billion years ago, life emerged. How it emerged is still uncertain, but the story of evolution begins with these first inhabitants of Earth. The span of time during which they emerged and first began to diversify into a number of forms is called the Precambrian (4.6 billion years to 543 million years ago). This vast age comprises nearly 90 percent of all Earth's history.

THE FIRST ORGANISMS

The first organisms emerged in the sea, and all living things, past and present, still bear their stamp. Our cells, like the cells of all organisms, require liquid water as the medium for circulation and chemical reactions. The precursors of the first organisms were probably organic molecules that spontaneously synthesized from simpler compounds containing carbon, nitrogen, oxygen, and hydrogen. Somehow, a transition was made from these molecules to things that could metabolize food and self-replicate, and were therefore

alive. But how that transition occurred remains unknown. It is not even clear where in the sea the first transitional steps took place. The first organisms may have emerged in the ocean's surface waters, where sunlight would have provided an energy source. Or, perhaps shielding itself from the Sun's ultraviolet

Above: A hydrothermal vent on the ocean floor spews super-hot, mineral-rich water. Although the sea bottom is populated sparsely, these vents provide an energy source for a vast array of life. Top left: An ocean wave. The first living things emerged from the sea.

radiation, life may have begun deeper down. It may even have emerged around hydrothermal vents, superheated openings in the ocean floor that still support communities that draw their energy from chemicals in the hot water rather than from sunlight.

The first living things known from the fossil record, from as early as 3.5 billion years ago, are bacteria. The first bacteria lived off organic compounds in the ocean water, but in time some evolved that had the capacity for the production of food from sunlight (photosynthesis).

THE SPREAD OF OXYGEN

Living things can unwittingly make drastic changes to their environment. Precambrian bacteria were no different. As they performed photosynthesis, they pumped oxygen as a by-product into the atmosphere. By about two billion years ago, atmospheric oxygen had accumulated in large quantities. The change was pivotal. Our present proportion of oxygen in the atmosphere— 21 percent—is vital to our survival and that of all other aerobic (oxygen-breathing) organisms.

Ironically, the bacteria that produced the oxygen did not need it and indeed found it poisonous. Some died off or found shelter in anaerobic

habitats such as deep mud. But others evolved into organisms that could safely use oxygen to release energy from food. This aerobic respiration is more efficient than the anaerobic kind, and it became the basis of many new forms of life.

EUKARYOTES

By 2.1 billion years ago, life included more structurally specialized organisms whose genetic material was contained inside a nucleus (eukaryotes). Bacteria, whose genetic material is not contained within a nucleus, are prokaryotes. The earliest eukaryotes were one celled, like bacteria. But many-celled eukaryotes evolved by 1.3 billion years ago, in the form of algae. Multicelled animals followed by about 600 million years ago, including some that reproduced sexually.

About 590 to 545 million years ago, the Ediacaran fauna lived. They included jellyfish, sponges,

Top: This diorama based on fossils from South Australia depicts Ediacaran fauna that first emerged during the Precambrian era. They were among the first multicelled organisms. They included jellyfish, sponges, and animals that attached themselves to the sea floor. Above: Present-day sea fauna at the bottom of a Scottish loch.

and animals attached to the seafloor like plants. They also included worms with a body cavity able to contain internal organs— a feature shared by most creatures that subsequently developed, including humans. None of the animals had mineralized hard parts, such as shells or skeletons. Those would come in the next era.

From Sea to Land

For nearly 300 million years after the Precambrian period, life evolved from relatively few multicellular forms to a bewildering array of animals and plants. Many of the familiar kinds of organisms we know today, from fish to trees, appeared during this long span of time. Other life forms, such as trilobites, have long since become extinct. This age was the Paleozoic era (543 to 248 million years ago), which included six periods: Cambrian, Ordovician, Silurian, Devonian, Carboniferous, and Permian. Among the momentous steps that living organisms took during this era was the transition from sea to land.

THE CAMBRIAN EXPLOSION

At the beginning of the Paleozoic came a rapid diversification of life known as the Cambrian explosion. Not only were there many new species of animals, but the Cambrian period (543 to 490 million years ago) was marked by the appearance of nearly all the phyla, or basic body plans, that exist today, along with some that no longer exist. The world of the Cambrian was a warmer and wetter one than today's world. Cambrian animals were all marine; no multicellular life existed on land.

Trilobites were hard-shelled marine animals that flourished during Cambrian period.

Cambrian animals differed from their predecessors in that they had hard parts, including shells, exoskeletons, and claws. Despite the diversity of hard parts, the backbone was not one of them: All Cambrian animals were invertebrates. The Cambrian and the period that followed, the Ordovician, are known as the Age of Invertebrates for the ascendancy of those animals.

Trilobite fossils comprise nearly three-quarters of all Cambrian fossils that have been discovered and identified. These hard-shelled marine animals were arthropods, members of the same phylum that today includes insects, spiders, and crustaceans. Other Cambrian animals

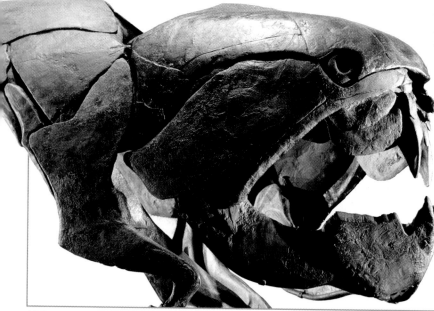

Fish appeared about 443 million years ago. Devonian-era fish, like this one, were the first to evolve scales. Top left: Horsetails were among the first seed-bearing plants.

Sea life during the mid-Cambrian period contained a variety of sponges, mollusks, echinoderms, and trilobites. These animals had exo-skeletons, shells, and claws, but no backbones. Vertebrates developed about 490 to 443 million years ago during the Ordovician period.

included sponges, mollusks, and echinoderms. But there were also animals that fit no known category; one example is *Opabinia*, which had odd appendages and numerous eyes.

THE AGE OF FISHES

During the Ordovician period (490 to 443 million years ago), vertebrates developed. Then, as now, these organisms featured a spinal column made of bone or cartilage. The first vertebrates were jawless, armor-plated fish called agnathans. In the Silurian period (443 to 417 million years ago), fish evolved jaws. In the Devonian period (417 to 354 million years ago), they evolved scales. Devonian fish included lungfish, armored fish, and sharks, with skeletons made of cartilage, and bony fishes such as ray-finned fish, which included

the ancestors of most modern fish. So great was the dominance of fish in the Devonian that this period is known as the Age of Fishes.

Even as fish evolved in the sea, great changes took place on land. The first land plants evolved in the Ordovician, and the first land animals—arthro-pods such as millipedes and mites—emerged in the Silurian. By the Devonian, insects had appeared and some fish had developed lungs, enabling them to live on land. Some of the latter evolved into the first amphibians, animals that live part of their lives in water and part on land.

THE FOREST PRIMEVAL

By the middle Devonian, the first trees had appeared. In the next period, the Carboniferous

(354 to 290 million years ago), these early organisms multiplied into vast, swampy forests that included the first seed-bearing plants. The vegetation, which included giant ferns, horsetails, forest trees, and club mosses (plants resembling large mosses), towered nearly 150 feet high. Winged insects and roaches flour-ished and the diversification of amphibians became widespread, giving this period the name Age of Amphibians.

Even in the Carboniferous, a rival to amphibians was emerg-ing: the reptiles, which could live on land throughout their lives. During the Permian period (290 to 248 million years ago), as the world climate became cooler and drier, reptiles spread. After the mass extinctions of invertebrates that ended the Permian, they became the dominant animal.

The Age of Dinosaurs

The best known of all prehistoric life forms were the dinosaurs. These reptiles included giants larger than any land animal before or since. But versions that were considerably smaller—about the size of a modern-day crow—were present as well.

Dinosaurs dominated Earth for most of the Mesozoic era (248 to 65 million years ago), a stretch of nearly 200 million years that encompassed three periods: the Triassic, Jurassic, and Cretaceous. The dinosaurs spread around the world, filling ecological niches from herbivore to carnivore to scavenger. At least some may have been warm-blooded. Ultimately, they became extinct, vanishing

from the biological stage as if they had never existed. Dinosaurs were not the only important players of this era.

Among the other kinds of organisms that evolved during the Mesozoic were two that are still with us: birds and mammals. New fossil evidence of turkey-sized ceatures called theropods reinforce the now widely held belief that the birds are descended from dinosaurs.

THE FIRST DINOSAURS

The first dinosaurs evolved about 230 million years ago, during the Triassic period (248 to 206 million years ago), the first of the Mesozoic's three periods.

Today's snakes are descended from diapsids.

The Triassic saw a continued drying of the global climate, which favored the spread of reptiles, which were better adapted for arid conditions than amphibians. During the Triassic, the formation of the supercontinent called Pangaea was completed, providing a global platform for the spread of new land animals.

One group of reptiles, known as the diapsids, included the ancestors of modern-day lizards and snakes. A separate group called the archosauromorphs gave rise to dinosaurs. Some of these, such as *Coelophysis*, were carnivores. Others were plant-eating, long-necked animals called prosauropods. The first mammals appeared in the Triassic.

THE JURASSIC HEYDAY

During the Jurassic period (206 to 144 million years ago), the climate grew rainier and warmer, with lush forests spreading and the sea level rising. The thick

Above: The Diplodocus *dinosaur was a long-necked herbivore. Top left: Ichthyosaurs, which resembled dolphins, were marine carnivorous reptiles of the Mesozoic era.*

vegetation spurred the evolution of plant eaters including sauropods. These gigantic four-legged animals reached dimensions of more than 100 feet long and 50 tons in weight. The largest land animals the world has ever known, they included *Apatosaurus* (formerly known as *Brontosaurus*), *Diplodocus*, and *Argentinosaurus*.

The enormous size of the sauropods would have given them advantages for both digestion and defense. Plant food is tough to digest, and herbivores need a large belly to store it in while the microbes in their gut decompose it. Size would also have been a defense against carnivores, which were evolving to take advantage of the feast represented by the new herbivores.

The new breed of hunters were theropods—lean, fast, two-legged dinosaurs with sharp teeth and claws. Large ones included *Allosaurus*, up to 33 feet long, but smaller ones—8 to 20 inches in length—were developing into a new kind of animal: the bird. By the late Jurassic, birds had emerged as animals covered with feathers and able to fly.

THE LAST DINOSAURS

The Cretaceous (144 to 65 million years ago) was the last period of the dinosaurs. It was extraordinarily warm, with forests growing and dinosaurs grazing in the Arctic and Antarctic circles. Many new varieties of dinosaurs evolved, especially ornithopods, plant eaters such as the hadrosaurs, or duck-billed dinosaurs. Among the new predators, the greatest

was *Tyrannosaurus rex*, which was 40 feet long and had six-inch teeth. Smaller carnivores included the sickle-clawed, birdlike hunter *Velociraptor*.

Angiosperms, or flowering plants, also evolved in the Cretaceous. They coevolved with insects that became specialized to carry their pollen from flower to flower, assisting their reproduction.

The Cretaceous came to an end with a tremendous mass extinction. The mammals, flowers, and insects survived, but the dinosaurs vanished.

The Velociraptor *was only two meters long and stood about one meter tall. This birdlike carnivore lived at the end of the dinosaur era.*

The fossilized skeleton of a plesiosaur, a marine reptile. The typical plesiosaur had a small head, a long neck, and limbs like paddles, which allowed it to navigate in water.

REPTILES OF LAND AND SEA

While dinosaurs dominated the land, other reptiles spread into the sea and air. Competing with birds for the Jurassic skies were the pterosaurs, reptiles with wing membranes that extended from the side of the body, along the arm, and were supported in part by a very long fourth finger. Pterosaurs included some of the largest flying animals of all time, with wingspans of up to 40 feet. *Pteranodon*, a large reptile with a crest at the back of its head, is one of the best-known pterosaurs.

In the sea, reptiles called ichthyosaurs and plesiosaurs flourished. The ichthyosaurs resembled dolphins, while the typical plesiosaur had a long neck, a small head, and paddlelike limbs. Legendary creatures such as Scotland's Loch Ness monster and the Ogopogo of British Columbia, Canada, if they indeed exist, may represent surviving plesiosaurs.

A New Kind of Mammal

After the dinosaurs died, a new interval began. The Cenozoic era, which commenced 65 million years ago, continues to the present day. During the Cenozoic era, mammals took over the role of dominant land vertebrates, giving the era its epithet the Age of Mammals. All modern species of mammal evolved during this era, while birds and flowering plants proliferated. One lineage of mammals gave rise to a new kind of primate: two legged, tool using, verbal, and intelligent. That primate was the human being.

The Cenozoic has two periods: the Tertiary (65 to 1.8 million years ago) and the Quaternary (1.8 million years ago to the present).

Above: The imperial mammoth was one of the species that helped fill the sizable biological void created by the dinosaurs' extinction. Top left: A depiction of the landscape and animals 33,000 years ago.

THE RISE OF MAMMALS

Evolved from reptilian ancestors, mammals existed side by side with dinosaurs since early in the Mesozoic era. During that time, mammals were small, furry animals, similar to rats and shrews. Originally egg layers, they evolved live birth by the end of the Mesozoic. They were otherwise unimpressive, active by night, eating insects, and living in burrows.

With the disappearance of the dinosaurs, mammals came into their own. They diversified in many directions, becoming larger and smarter, filling the niches that the dinosaurs had abandoned. Ancestors of many modern groups appeared, from rabbits to cattle to elephants. During the Eocene epoch (55 to 34 million years ago), the first bats took to the sky, whales proliferated in the seas, and the primates called monkeys evolved in tropical forests. In the Oligocene epoch (34 to 24 million years ago), ape primates appeared in Africa.

BIRDS AND FLOWERS

Despite the prominence of mammals, other groups of living things were also flourishing. During the Paleocene epoch (65 to 55 million years ago, the first epoch of the Cenozoic era), carnivorous mammals were outnumbered by birds. These included large, flightless predators such as *Diatryma*, which stood six feet tall and had a powerful beak and sharp claws. Also during the Paleocene, flowering plants (angiosperms) became the dominant land plants.

THE EVOLUTION OF HUMANS

About eight to five million years ago in Africa, a new kind of mammal split off from the rest of the ape family. This was the hominid (humanlike) branch, the line from which modern humans evolved. Unlike their tree-dwelling ape ancestors, hominids walked on two legs; they were bipedal. They may have left the trees in response to a drying of the global climate, which reduced forestation and spawned more grassland.

About four million years ago, a group of hominids called australopithecines evolved. They had apelike faces and brains about one-third the size of a present-day human's. The first advanced glimmers of intelligence occurred about 2.5 million years ago as hominids with larger brains emerged. Members of a new genus (our own, *Homo*), they included *Homo habilis*, which had sophisticated stone tools. Around 1.8 million years ago, *Homo erectus* evolved. Brain development was even more refined, perhaps even

to the point of language. This species moved out of Africa and colonized parts of Europe and Asia. Other hominids also moved out of Africa, including the squat, massive Neanderthals.

Modern humans, *Homo sapiens*, evolved by about 100,000 years ago. How this species originated is still uncertain. Some scientists hold that all humans alive today are descendants of one African population (the "out of Africa" view), while others hold that hominids in many parts of the world evolved separately but in step to yield the current human species (the multiregional view). Whichever account is true, only *Homo sapiens* survived the ice ages of the Pleistocene epoch (1.8 million to 10,000 years ago) to become the sole hominid species.

Top: About 2.5 million years ago, a more sophisticated group of hominids emerged. They were capable of fashioning tools like these stone axes. Below: This Homo ergaster *skull was discovered in Kenya and is about 1.7 million years old. Bottom left:* Homo erectus *evolved about 1.8 million years ago.*

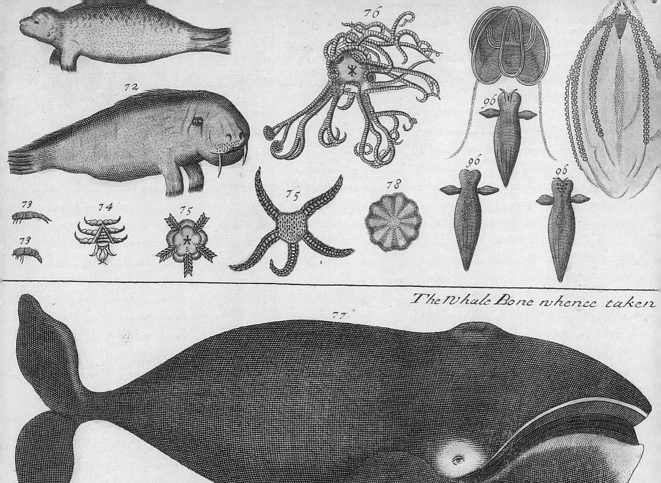

72

76

96

96

96

73
73
74
75
75
78

The Whale Bone whence taken

77

Another Whale
77
79
79
79
85
83

The Fin Fish.
95

CHAPTER 9

WHAT'S IN A NAME?

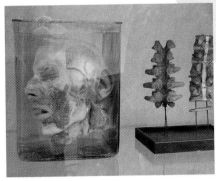

Left: A variety of marine life, including the walrus (72), and fin whale (95). Taxonomy, the science of naming and classifying living things, clarifies evolutionary relationships. It shows how animal groups are descended from ancestral groups. The whale and walrus are both mammals and occupy the same kingdom, phylum, and class, but they differ in order. The fin whale belongs to the kingdom Animalia, *phylum* Chordata, *class* Mammalia, *order* Cetacea. *The walrus, however, belongs to the order* Carnivora. *Top: A school of fish. Bottom: Exhibition items in an autopsy room, including a human head.*

Life on Earth consists of more than 10 million species, from bacteria to whales. Describing each species as if it were an independent entity would be a difficult task and would ignore the significant similarities with other species. A more reasonable approach to studying the planet's biodiversity is to group species meaningfully and research shared and distinct traits. For centuries, this effort has been the work of taxonomists—scientists who name and classify living organisms.

As a practical matter, this classification and organization is extremely important. For a nonbiological example, consider the organization of a kitchen. If things were organized alphabetically, or not at all, finding what one needs would be extremely difficult. Similarly, taxonomy creates an easier way to get around the biosphere. It also clarifies evolutionary relationships, showing which groups have likely descended from various ancestral groups.

The modern classification system uses a hierarchy of groups, from kingdom at the top down through phylum, class, order, family, genus, and species. The kingdoms are often considered to be: prokaryotes, protists, fungi, plants, and animals, with some variations. This chapter discusses prokaryotes, protists, and fungi.

Systems of Order

Classifying and organizing life are age-old endeavors. In ancient Greece, the philosopher Aristotle grouped living things according to similarity. In the intervening millennia, classification systems have progressed as living things have become more thoroughly understood.

Classification is not always self-evident, however. Visualize a bedroom closet, for example. One person might organize the clothing within by color, whereas another might categorize based on other criteria, from summer to winter, perhaps, or casual to formal. People might also disagree on what goes into the closet in the first place: Do shoes belong in the bedroom closet or the back hall? Taxonomists agree on many groupings and criteria, but it is important to keep in mind the differences and discrepancies between systems

ALTERNATIVE SYSTEMS

Our current classification system is based on that developed by Carolus Linnaeus, an eighteenth-century Swedish botanist who grouped organisms by structural similarity. He used a hierarchy of groups, with the largest group being the most inclusive (all animals, for example) and the smallest group the least inclusive (all African bush elephants, for example). Many taxonomists have contributed to the improvement of his system by incorporating an understanding of evolutionary relationships. But some Linnaean features are still in use, such as the identification of groups of organisms by Latin names.

Aristotle's taxonomic approach to classification of organisms divided animals and plants into broad categories. Plants, for instance, were subdivided according to their shape and into broad categories like trees, shrubs, and herbs.

Above: Swedish botanist Carolus Linnaeus organized life into a hierarchy of groups. The African bush elephant belongs to the least inclusive grouping. Top left: Pleurosigma angulatum, an abundant diatom species.

Biological classification is constantly changing. A biology text from the early 1900s typically divided all life into just plants and animals. Contemporary textbooks often list five kingdoms: prokaryotes (Prokaryotae), protists (Protista), fungi (Fungi), plants (Plantae), and animals (Animalia).

With respect to classification, though, complete agreement does not exist. Since the 1980s, some scientists have divided life into three domains: eukaryotes (which include the protist, fungus, plant, and animal kingdoms), archaebacteria (which include bacteria that live in extreme environments), and eubacteria (which include more common bacteria). Proponents of this system emphasize that the three domains represent ancient evolutionary lineages overlooked by the five-kingdom system. Another approach recognizes six kingdoms: protists, fungi, plants, and animals, but with prokaryotes divided into two distinct kingdoms, archaebacteria and eubacteria.

VIRUSES AND PRIONS

Viruses and prions are not so easily characterized. They are generally not even considered to be alive, since they are not capable of independent reproduction. But they interact with living things in ways that seem living.

Viruses, which include human immunodeficiency virus (HIV, the cause of acquired immunodeficiency syndrome, or AIDS) and influenza (the cause of the flu), are tiny packages of DNA or RNA that, while lifeless outside of a host, co-opt the cells of host organisms and force them to reproduce more viruses.

Prions, the cause of bovine spongiform encephalopathy (BSE, also known popularly as mad cow disease), are proteins that are a normal constituent of humans and other creatures. In their normal three-dimensional shape,

they have an as-yet unknown function. The problem comes when, for reasons unknown, they adopt a different conformation. This shape-shift triggers a conformation change in an adjacent prion, and so on, which progressively damages cells in the brain. The result is a progressive loss of function. This effect occurs purely because of the protein's shape and does not involve gene activity. Neither viruses nor prions are composed of cells, nor have the capacity to reproduce outside a host. They are therefore not usually included as part of the biological classification system.

A multitude of life inhabits the microscopic world, as on this laboratory slide.

MAD COW DISEASE

Mad cow disease became big news in the 1990s with an outbreak in England. About 200,000 cows were infected, and at least 140 Britons acquired the human equivalent, Creutzfeldt-Jakob disease (CJD). More recently, diseased cows have been identified in the Canadian and U.S. cattle industries, although they did not enter the food supply of either nation. The progressive brain cell damage caused by prions is evident as changes in behavior and coordination. About 90 percent of people infected die within a year of diagnosis. Infections mainly occur through diet. Cows are often fed the otherwise unusable remains of other butchered cows. When that feed contains infected nervous tissue, cows can become ill. Humans can contract the disease by eating meat contaminated with pieces of cow brain or spinal tissue.

Prions are difficult to categorize because they are incapable of reproducing outside of a host. They are the cause of mad cow disease, becoming destructive when their normal shape changes.

The Categories of Life

Carolus Linnaeus is often considered the father of taxonomy. The principles he developed for classifying living things hierarchically still permeate the field today.

Linnaeus grouped similar species into a larger group, called a genus. Then, he grouped similar genera into families, and so on. In all, he designated seven levels: kingdom, phylum, class, order, genus, and species. At the species level, there are a few species that share many traits. In a kingdom, there are many species that share a few traits. A common mnemonic device to remember the groups, from largest to smallest, is: King Phillip Came Over For Great Spaghetti.

THE LINNAEAN SYSTEM

In the Linnaean system, every living thing is identified by a two-part name. The first part of the name is for the genus, or group. The second part is for the species, or kind. Humans belong to the genus *Homo* and the species *sapiens*. By convention, taxonomists capitalize the genus name and lowercase the species name, writing the whole name in italics: *Homo sapiens*. Humans belong to the kingdom Animalia, phylum Chordata, class Mammalia, order Primates, family Hominidae, genus *Homo*, and species *sapiens*.

Dogs also are in the kingdom Animalia, phylum Chordata, and class Mammalia,

Carolus Linnaeus is considered the father of modern taxonomy. His treatise on taxonomy, Systema Naturae *(1758), classified 4,400 species of animals and 7,700 species of plants. His Latin naming system gave every organism a two-part name.*

Above: A crayfish with small mussels on its hull. The Linnaean system helps scientists to distinguish among similar species within a family or genus, such as the hundreds of crayfish species. Top left: Deer, which belong to the class Mammalia, are descended from reptiles.

but their taxonomic similarity to humans ends there. They are in the order Carnivora, family Canidae, genus *Canis*, and species *lupus*. Dogs share some characteristics with humans, such as vertebrae and hair, but stark differences exist as well. These similarities and differences are reflected in their taxonomic classification.

PROBLEMS WITH ORDER

Taxonomy is tremendously convenient for scientists. Scientific names help researchers to remember the traits of groups and to communicate about species with the confidence that they are referring to the same living things. Common names will not do the trick. For example, there are hundreds of crayfish and crawdad species in North America; simply using their common names does not help distinguish all the different species among them.

But convenience can lead to problems. In traditional classification systems, living things are grouped according to similarity. To decide which organisms are most similar, traits such as body length, presence of wings, number of legs, and reproductive capacity are chosen. Organisms are then grouped by the number of traits they share. One problem with this approach is how to determine which similarities should be considered, since a potentially infinite number of traits exist. A

This Madagascar Moon Moth belongs to the genus Argema, *species* mittrei. *Hence, this moth is known as* Argema mittrei. *The* Argema *genus is classified according to its geographic location: eastern Africa.*

second problem is that many taxonomists consider the process an unjustifiable human construct. It seems as if scientists arbitrarily create categories to decide which organisms belong in which groups.

EVOLUTIONARY CLASSIFICATION

To deal with these objections, most taxonomists favor a system that combines Linnaean principles with classification based on evolutionary relationships. Many similarities between living things, including some of the criteria used in traditional classifications, are the result of relatedness. In a

sense, the evolutionary system has been overlaid on the hierarchical system to create the modern classification system used today.

Contemporary taxonomy is meaningful because it shows how organisms are related. Taxonomic relatives are kinship groups that share a common ancestry—mammals evolved from reptiles, for example. To describe deer as mammals is not to put them into an arbitrary category, but to delineate how they evolved from reptiles.

The Family Tree of Organisms

The traditional Linnaean classification system reigned supreme for about 200 years. But in the 1950s, German taxonomist Willi Hennig proposed infusing more biological meaning into classification. Instead of simply choosing arbitrary traits, he argued that classification should be based on evolutionary relatedness. With the discovery of DNA and development of improved molecular techniques, scientists have employed molecular traits to sort out these relationships.

Some traits are shared among species because their common ancestor had those same traits; examples include the hair found on mammals or the characteristic four legs of tetrapods. Only the traits arising from a common ancestor are considered when

Above: A robin. Top left: Black Caiman of South America. In the 1950s, the German taxonomist Willi Hennig paved the way for a system that clarified evolutionary relations among species, like the robin and alligator.

determining who is related to whom. Other characteristics, such as the streamlined body shapes of tuna and dolphins, might appear similar, but they have actually evolved multiple times in separate lineages and are therefore not useful in determining relationships.

PHYLOGENETIC TREES

Illustrations called phylogenetic trees can help visualize these relationships. The tips of the branches represent living species, and species originating from the same main branch share similar traits. Species from opposite sides of the tree might share a few traits and be in the same large, inclusive group. But species on nearby, adjoining branches would share many more traits. Each branching point represents a shared ancestor.

For example, the trunk of the tree could represent primitive vertebrates no longer found on Earth. Therefore, all the animals on this tree would be vertebrates. Moving up the tree, the first branch and all of its twigs would represent fish, the next branch up the main trunk would represent amphibians, followed by reptiles then mammals, and then the rest of the vertebrates. Each branch is a group of animals that share traits that

distinguish them from other groups and that have evolutionary relevance. Each branching point represents a shared ancestor between the two lineages.

These trees can be used for classification. Each group in the diagram contains species that share a common ancestor. This taxonomic method is called cladistics.

CLADISTICS AND TRADITION

In many cases, classifications based on cladistics and on similarity result in the same groupings. Examples of inconsistencies do exist, however. Superficially, a consideration of dolphins, trout, and elephants could suggest that dolphins and trout belong together in one group, separate from elephants. In fact, elephants and dolphins are much more closely related (they are both mammals), once evolutionarily relevant traits are factored in. Likewise, traditional classification has historically grouped crocodiles, lizards, and snakes as reptiles. But a cladistics analysis clusters birds and crocodiles as one group, and snakes and lizards as another.

With millions of species inhabiting the Earth, a nearly infinite number of hierarchical groups could exist. Therefore,

Above, from left: Tuna and dolphins share a streamlined body trait. But they have evolved multiple times in separate lineages, the dolphin belonging to the class Mammalia, the tuna to the class Actinopterygii. Their body traits, therefore, are not useful in determining relationship because they belong to different classes. The dolphin belongs to the class Mammalia while the tuna belongs to the class Osteichthyes, or bony fishes.

a strictly cladistic system would be impractical. Scientists have instead settled on a combination of the traditional and cladistic methods. Certain branching points in the tree of life correspond to kingdom, phylum, class, and so on. Sometimes additional levels are created, such as subphylum or subclass.

Advocates of the combined system argue that it is a natural arrangement, not a human invention, because it corresponds to evolutionary relationships. As these relationships are untangled, classification will continue to improve.

WHY IS A SHARK UNLIKE A SARDINE?

Sharks and sardines are both fish—but one strikes fear into ocean goers, while the other sits canned on supermarket shelves.

The two fish are strikingly different. Sharks—despite their toughness—lack bone. Only cartilage supports their bodies. Their fins are fleshy and they breathe through gill slits. Unlike sardines, they add scales to their bodies as they grow. Sardines, on the other hand, have true bone. Their fins are thin and are supported by rays. Moreover, their gills are covered by a bony plate. Their scales grow as they grow, marking time like tree rings.

Why are there so many differences? Because their most recent common ancestor lived 370 million years ago, during the middle Devonian period. At that time, the lineage of sharks, with their cartilaginous skeletons, split off from that of the bony fishes, including sardines.

Unlike sardines (corner), skeletal frames of sharks consist of cartilage. They had a common ancestor 370 million years ago.

The Oldest Living Things

Of the five kingdoms of living things—prokaryotes, protists, fungi, plants, and animals—the oldest are the prokaryotes. Prokaryotes, as typically represented by bacteria, are simple in structure but complex in function. They can move toward nutrients and resources, recoil from toxins, sense up and down, thrive in extreme environments, and even form predatory colonies.

The effects of prokaryotes on human life are many. Bacteria cause disease, provide antibiotics, and make possible such foods as pickles, kimchi, sauerkraut, yogurt, and buttermilk. In the human gut, friendly bacteria outnumber human body cells. They are ancient as well as successful, and abundant.

BACTERIAL STRUCTURE

Bacteria differ from eukaryotes in that they lack internal compartments bound by membranes. That means they lack a nucleus, a mitochondria, an endoplasmic reticulum, and chloroplasts. Despite the absence of these structures, bacteria can carry out the processes of life and are amazingly diverse.

Each bacterium has DNA in its single cell. The genetic material is often apportioned as a main chromosome and smaller circles of auxiliary DNA (plasmids).

Above: Yogurt. Top left: Antibiotic medication. The good types of bacteria make yogurt and antibiotics possible.

Bacteria also contain ribosomes, which, as in eukaryotes, are structures that aid in protein synthesis. Bacteria are surrounded by a cell membrane, and some have an additional cell wall that can help them attach to surfaces ranging from rocks in a riverbed to teeth in a human mouth. Attached bacteria are also often bunched together in a slimy layer; this organized bacterial community is known as a biofilm. Some bacteria have photosynthetic pigments, although they are not contained in chloroplasts. Like eukaryotes, they have a whole host of enzymes to carry out chemical reactions.

Bacteria can be round, such as *Streptococcus mutans*, which causes tooth decay. Others, like the food-poisoning agent *Bacillus cereus*, are cylindrical. Still others are squiggly, such as *Treponema pallidum*, which causes syphilis. Many have tail-like flagella that propel them around, or are covered with shorter, hairlike structures that also aid in movement. Other hairlike structures, called pili, function in the transfer of genetic material from one bacterium to another. Bacteria reproduce by splitting, in a process called fission, and some do so every 10 minutes. They can also "mate" by joining cell membranes, swapping plasmid DNA, and then separating. This genetic exchange, together with frequent fission, contributes to the rapid rate of evolution observed in bacteria.

TYPES OF PROKARYOTES

Prokaryotes are divided into two groups: archaebacteria and eubacteria. Archaebacteria live in extremely harsh conditions. Their members include methanogens, which live in the acidic stomachs of cattle, where they aid in the cows' digestion, which in turn releases methane into the atmosphere as a by-product. Halophiles (salt lovers) are another type of archaebacteria. They can be found in salty

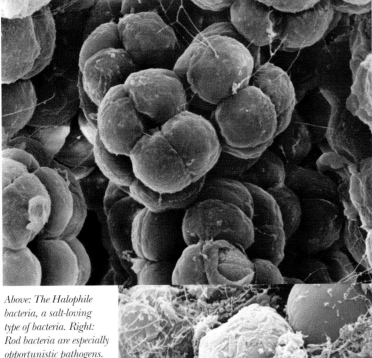

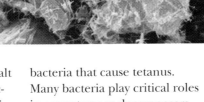

Above: The Halophile bacteria, a salt-loving type of bacteria. Right: Rod bacteria are especially opportunistic pathogens. They attach to mucosa cells in the human intestine. Prokary are divided into two types of bacteria: Archaebacteria and eubacteria. The Archaebacteria like Halophile and Rod bacteria can thrive in extreme conditions.

Regular hand washing is more than a way to maintain healthy personal hygiene. Properly done, hand washing is a line of defense against the spread of disease.

WASH YOUR HANDS

Bacteria are everywhere—on doorknobs, computer keyboards, elevator buttons, shopping carts—and many of them can make people sick. Catching a cold is as simple as touching a contaminated surface and then rubbing one's eyes or nose.

Once a person gets sick, everything he or she touches can spread the bacteria to other people. And the danger is not just the common cold—the microorganisms that cause hepatitis A, meningitis, and infectious diarrhea, among other illnesses, can be spread this way.

Hand washing is a simple practice that can halt the spread of disease. But the hand washing must be done properly. The Centers for Disease Control recommends wetting your hands and applying soap. Next, rub vigorously for 10 to 15 seconds. Try humming the tune to "Happy Birthday" instead of counting. Finally, rinse and dry. Antibacterial soap is not necessary. Washing with regular soap does the same job and does not contribute to antibacterial resistance.

environments such as Great Salt Lake, Utah, among other places, but they thrive in even saltier conditions. Archaebacteria also inhabit extremely hot environments. Hot springs such as those in Yellowstone National Park harbor these thermophiles (heat lovers). Archaebacteria are more closely related to eukaryotes than eubacteria are.

Eubacteria encompass the rest of the bacteria, including blue-green algae (nitrogen fixers that act as nature's fertilizer) and

bacteria that cause tetanus. Many bacteria play critical roles in ecosystems as decomposers.

Some bacteria, such as the tetanus-causing *Clostridium tetani,* can form endospores— essentially a clonal offspring in a protective casing—when resources are scarce. Endospores withstand heat, dessication, acids, and disinfectants and remain dormant until favorable conditions return. Endospores can successfully germinate even after a century of dormancy.

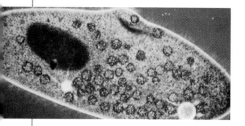

One-Celled Wonders

Tiny organisms that are not bacteria also permeate the environment. They are eukaryotes, complete with nuclei and other cellular compartments, but they are difficult to classify among other eukaryotes. Most are one-celled. Protists, as they are called, range from malaria parasites to sea lettuce, and are broken down into three groups: fungi-like, animal-like, and plantlike.

FUNGI-LIKE PROTISTS

Protists that resemble fungi include predatory and parasitic molds, which are decomposers in fresh water and marine habitats. Like fungi and some bacteria, they produce spores and absorb nutrients. But, unlike fungi, they actively engulf food and can aggregate and migrate to their equivalent of greener pastures when nutrients are scarce. Downy mildew seen on grapes and fuzzy mold found on aquarium fish fall into this category.

The most famous of the parasitic molds is *Phytophthora infestans*, the protist responsible for the Irish potato famine of 1845–50. Several years of damp growing seasons contributed to the protist's population explosion, and one-third of Ireland's population starved as a result. Even today, *Phytophthora infestans*

is considered the single most costly biological constraint on global food production.

ANIMAL-LIKE PROTISTS

Animal-like protists can be predators, grazers, or parasites. They have complex life cycles and actively move through the environment. Like the other protist groups, they are quite diverse. Some species house photosynthetic algae within, while others are free-living. The amoeba and the paramecium, members of a group called protozoa, are animal-like protists.

Some animal-like protists are major pathogens, such as Plasmodium, the protozoan that causes malaria. Giardia lamblia,

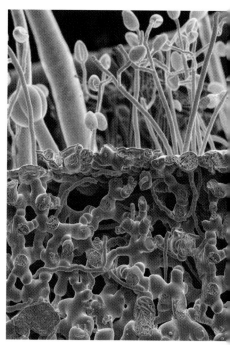

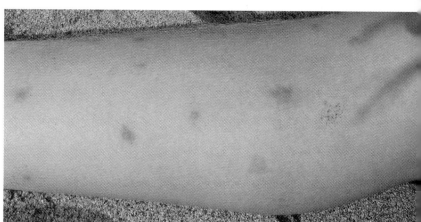

Top: Electron microscope image of potato blight fungus. Bottom: Arm riddled with mosquito bites. Malaria, which is transmitted by mosquitoes, is caused by an animal-like parasite. The fungi-like potato blight and malaria parasite are two of the three types of one-celled organisms called protists. Top left: The Paramecium protozoan *is an animal-like protist.*

a species known to campers as the cause of "beaver fever," is found in freshwater streams and can cause weeks of intestinal discomfort and diarrhea if untreated water is ingested. Another protist causes African sleeping sickness. But not all animal-like protists cause such human suffering. Foraminifera created the English white cliffs of Dover, with the fossilization of their shells 200 million years ago.

PLANTLIKE PROTISTS

Plantlike protists are critical to aquatic food webs: Using photosynthesis, they convert sunlight energy into usable forms for other organisms. Without them, aquatic food webs would collapse. These protists can be single-celled or can form colonies; they either drift or swim through the water.

Dinoflagellates are one group of plantlike protists, some of which kill billions of fish during red tides, episodes in which large numbers of microorganisms turn water reddish; other dinoflagellates create beautiful displays of bioluminescence. Red algae form colonies and protect themselves with slippery mucus. Red algae is probably best known to landlubbers as nori, the sushi wrapping.

Kelp is a form of brown algae, another example of a plantlike protist. Kelp forests support huge underwater communities, where fish, abalone, lobsters, and other organisms live. Products that contain kelp compounds include ice cream, jelly beans, salad dressing, toothpaste, and paper. Green

Kelp, a plantlike protist, plays a key role in the aquatic food web. Through photosynthesis, aquatic protists can convert sunlight energy into usable forms for other organisms.

algae are other types of plantlike protists. In fact, they exhibit so many "plantlike" qualities that some taxonomists consider them plants. Structurally and biochemically, green algae are, indeed, very similar to plants. Some examples include edible sea lettuce and red algae, which turns snow pink or red.

A fungi-like protist is responsible for decomposing this apple and turning it bad. A similiar protist was responsible for devastating the Irish potato crop, causing famine in 1845–50.

Mushrooms and Mold

Many people associate fungi with mushrooms in the supermarket or the frightening food forgotten in the back of the fridge. But mushrooms and molds are much more than those simple images might suggest. In addition, mold is an ambiguous term, sometimes used to describe distantly related members of other kingdoms.

Fungi are heterotrophs, which means that they cannot produce their own food like plants can. Some are saprobes, decomposers that secrete enzymes to break down food into smaller components, which they then absorb. Other fungi, like those that cause athlete's foot, are parasites that live and feed on their hosts. Some are symbionts, living in concert with their hosts. Many are multicellular. The first fungi evolved 900 million years ago, and by 330 million years ago, the major groups alive today had been established. Millions of years of evolution have created vast diversity within the group.

GROUPS OF FUNGI
One group of fungi is Zygomycota, which includes terrestrial molds and soil decomposers such as black bread mold. One zygomycete inhabiting cow dung flings its spores in the direction of the sunlight. This action probably serves to propel the offspring onto fresh grass. Cows are likely to graze on the grass and therefore swallow the fungus, helping to complete its complex life cycle.

Ascomycota is a much larger group than Zygomycota, encompassing about three-quarters of fungi species, including yeasts and truffles. The unifying feature of species within this group is a specialized reproductive structure. To humans, yeasts are both good and bad—aiding in the production of bread and beer, but also causing athlete's foot and thrush. One species, *Claviceps purpure*, infects rye and other grains and produces toxins that stay active even after baking. The toxins cause ergotism, a poisoning that results in convulsions. Ergotism may have been the underpinning of the Salem witch trials of the American colonies, because the poison causes people to act abnormally and, as perceived at the time, possessed. Ergotism may have also led to the downfall of Russia's Peter the Great, whose soldiers and horses ate rye and went into convulsions. But before wholeheartedly condemning *Claviceps purpure*, keep in mind

Above: A brewery needs yeast to produce beer. Yeasts, a type of fungi, can be both good and bad for humans. A variety of yeasts aid in the production of beer and bread, but some species can produce harmful toxins. Top left: Shelf fungus growing on a tree stump.

that its by-products have been used to treat migraines, lower high blood pressure, and prevent hemorrhaging after childbirth.

The next group, Basidiomycota, includes the portobello. Coral fungus and the shelf fungus often seen on rotting logs in forests are also prime examples. One hallucinogenic species, *Amanita muscaria*, has been used in rituals in Central America, Russia, and India. Ingest too much, however, and it is fatal. Less familiar basidiomycetes are rusts and smuts, plant diseases that can wreak havoc on crops and are the bane of farmers.

MUSHROOM STRUCTURE

An umbrella-shaped mushroom seen aboveground represents just part of a whole fungus, the portion known as the fruiting body. Long filaments left behind after a mushroom is harvested remain underground and generate these reproductive structures.

The mycelium is the mushroom segment that remains underneath the soil. This segment may live and grow for many years. The fruiting body lasts only a few days, during which it produces reproductive cells called spores from which new mushrooms grow.

Farm-stand mushrooms. Black and white truffle mushrooms are a rare delicacy.

A FUNGAL DELICACY

Culinary connoisseurs covet the truffle, an underground fungal delicacy found mainly in Europe. Not to be confused with its chocolate counterpart, truffles are hunted by dogs and pigs on carefully guarded breeding grounds. Both pungent and expensive, truffles often fetch prices of hundreds of dollars per pound. They are served as thin slices on meat and as components of foie gras and pâtés. Their flavor has been compared to earthy gorgonzola cheese. Both black and white truffles are highly prized. Those on a budget can try the less expensive Chinese truffle or truffle-infused oil.

Above left: Basidiomycetes like corn smut can wreak havoc on crops. Top: Hallucinogenic Aminita muscaria *mushrooms. Bottom: Button mushrooms. The long filaments of mushrooms remain beneath the surface soil, generating reproductive structures.*

CHAPTER 10

PLANTS AND ANIMALS

Left: The domestic cat, one of the animals most familiar to humans, shares some of our anatomical and physiological traits. We rely on them for companionship and rodent control. The cat has proliferated through selective breeding by humans. Above: Insects such as the locust (top) have been able to thrive through interaction with humans. Swarms of locusts can decimate cultivated crops such as wheat (above), wreaking havoc and threatening human livelihood.

Plants and animals are the organisms in the biosphere with which humans are most familiar. They are classified through long-established methods that distinguish organisms broadly, such as by food or habitat, and also by increasingly specified shared characteristics, such as body structure for animals and leaf shape for plants.

In addition to being living organisms dependent on sun and water for sustenance, animals have much in common with plants. Both have regular daily and seasonal rhythms. Both are multicellular organisms with complex anatomies and physiologies. Both fall into regular classifications that include phyla, classes, and other categories that reflect their differences. Both exhibit adaptations to their surroundings.

Humans depend on plants and animals for food, tools, work, and companionship. In some cases, interaction with humans has caused animals great hardship and forced them to develop new adaptations. Overall, humans have had great effects on plants and animals, by increasing their number through domestication (in animals such as the cat and dog), by diminishing their number through overdomestication, or by thinning their ranks and endangering entire species.

Alike yet Different

Plants and animals have much in common. They are multicellular organisms that are adapted to live in many environments. They require nutrition and water to live. For the process of respiration, both need oxygen and release carbon dioxide. They grow from one cell to complex forms through stages of development that span an organism's youth and maturity. These stages are marked by changes in appearance or behavior. Through one cause or another, all plants and animals die.

Animals can cooperate and compete with other species and members of their own kind. So can plants: They engage in various types of cooperation and competition with other plants and animals to promote seed dispersal and reduce invading species.

DIFFERENCES

Plants and animals also have many differences. For example, plants account for the greater part of the world's biomass but have many fewer known species than do animals. There are approximately 300,000 known species of plants, while there are two million known species of animals. In addition, plants and animals get their nutrition differently. Plants obtain energy directly from sunlight through photosynthesis, and they use this energy to build organic matter. Animals get their energy from other organisms—living or nonliving animals or plants.

The cell structure of plants and animals is markedly different. In the case of a plant cell, the rigid cell wall consists of cellulose, a material not present in animal cells. Plants and animals differ in their mobility. Animals are the fastest organisms on Earth, able to react and move in response to their environment. They can do so because of their specialized muscular and nervous tissue. The nervous tissue receives data from the environment and sends appropriate signals to stimulate the muscles; the muscles move in response. In contrast, plants are so sedentary they are anchored to the ground with roots, which secure the plant and obtain water and minerals from the soil.

Animals are able to modify their behavior in response to past experience. This is not the case with plants, although some plants do have specialized parts, such as leaves shaped into spines, to ward off predators.

Life spans form another area of difference. The life span of an insect can be a number of hours; small animals such as rodents or birds may last weeks or months. Medium or large mammals live between 20 and 30 years. Humans can live longer than 100 years. The life span of plants, however, has an even wider range. Some may last several days or weeks, but some such as redwood trees can live for thousands of years.

Above: The common liverwort, a nonflowering plant, belongs to one of the three Bryophyte phyla. The others are mosses and hornworts. Top left: Moss lines rocks on a stream.

The giant sequoia is the world's largest tree, reaching up to 307 feet tall (93.6 m) and diameters of up to 29 feet (8.85 m).

CLASSIFICATION

Plants and animals are classified according to different criteria. Plants are classified in two basic ways: as flowering or non-flowering varieties, or as vascular (containing circulatory vessels) or nonvascular varieties. Bryophytes are three phyla of nonvascular plants, including mosses, liverworts, and hornworts. Tracheophytes are vascular plants that have vascular tissue (xylem and phloem) that delivers water and minerals from the roots to the stems and leaves and delivers food from the leaves to stems and roots.

Animals are classified by internal anatomy (vertebrates and invertebrates, cold-blooded and warm-blooded), habitat (land and water), food (herbivores and carnivores), patterns of development (breathing, loco-motion), and genetic makeup. There are about 30 phyla in the animal kingdom.

The study of animals is zoology; the study of plants is botany.

Desert plants, called xerophytes, often grow only when there is enough moisture in the soil to sustain life.

DESERT PLANTS

Many desert plants have evolved adaptations that enable them to survive in an environment with little water. These plants are called xerophytes.

Some xerophytes, such as desert annuals, do not have structural adaptations for conserving water. Instead, they avoid periods of drought by having their life cycle during the short period when there is enough moisture in the soil to sustain them. Plants that are active during the dry period have special adaptations in their leaves or roots. For example, some plant leaves have a protective epidermal layer such as a cuticle or epidermal hairs. A different adaptation occurs in the ocotillo, a plant common to Mexico and the southwestern United States. It produces leaves only when the region has sufficient water.

Plants with root adaptations include cacti, which have large, shallow root systems that trap water even during light rains. Some trees, such as a genus of the mesquite tree, have long taproots that can reach water sources located a considerable distance underground.

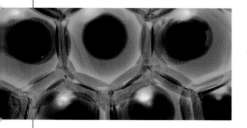

The Kingdom of Plants

Plants are multicellular organisms that make up the kingdom Plantae. They are characterized by their development from an embryo, presence of chlorophyll, and ability to carry out photosynthesis. Among the 260,000 or more species of plants are bushes, ferns, flowers, liverworts, mosses, trees, and vines. The bulk of these species are land plants: plants that rise from the ground. They evolved about 450 million years ago from green algae that existed around lakes and oceans and had cell structure, cell walls, and chlorophyll similar to plants.

The environment reaps benefits from plants in many ways, including erosion control, shelter, clothing, and medicine. Above all, they are a source of food.

FEEDING THEMSELVES

Unlike animals, which take in food from living or dead organic matter, plants make their own food through photosynthesis. Photosynthesis is the process of taking light energy from the Sun and turning it into chemical energy. In the green organelles called chloroplasts (which contain chlorophyll), plants use sunlight to make their own nutrition.

Fungi were once considered part of the plant kingdom but are no longer so classified because they do not have chlorophyll and their cell walls contain chitin instead of cellulose. They do not make their own food but acquire it from living or dead organic matter.

VASCULAR AND NONVASCULAR PLANTS

Plants may be divided into two distinct classifications: vascular and nonvascular plants. Nonvascular plants (bryophytes)

Above: Animals, like these cattle grazing in a pasture in California, take in food from organic matter. Plants make their own food through photosynthesis, or the conversion of light energy from the sun into chemical energy. Top left: Photo-magnification of plant cells.

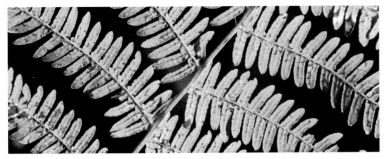
Ferns are an example of seedless vascular plants. While they have tissue that transports water and minerals from the soil, they reproduce using spores, not seeds.

have no leaves, stems, or roots. They also do not have internal vessels or leaf veins. In contrast to most vascular plants, nonvascular plants grow close to the ground. There are three phyla of nonvascular plants and about 16,000 species within them; they include mosses and other moss-like plants, such as the liverwort and hornwort. Because these plants lack roots or a transport system of vessels, they live and grow in pools of water.

Vascular plants (tracheophytes) account for the majority of plant species. Their name refers to the tubes that branch out through the plant and act as a nutritional delivery system. They have two types of vascular tissue: xylem, which transports water and minerals from the ground to the stem and leaves, and phloem, which transports food from the leaves to the roots, stems, and reproductive organs. Due to this delivery system, vascular plants often have strong stems and are taller than nonvascular plants. Vascular tracings are particularly distinctive on leaves.

Vascular plants can be further distinguished by the presence or absence of seed production.

Aside from seedless vascular plants (most of which are ferns or are related to ferns), vascular plants reproduce through seeds.

GYMNOSPERMS AND ANGIOSPERMS

Varieties of seeded vascular plants are gymnosperms and angiosperms. Gymnosperms are vascular plants that have seeds that are exposed, not enclosed in and protected with fruits. Fossil gymnosperms date back 350 million years. Varieties include conifers and similar plants such as cycads and ginkgoes. Their seeds develop in cones.

Angiosperms are vascular flowering plants whose seeds are protected inside an ovary and eventually ripen into a fruit. More complex than gymnosperms, angiosperms probably appeared on Earth later than gymnosperms, during the Mesozoic era (248 to 65 million years ago). Eventually they took dominance and now account for most plants and trees. Angiosperms reproduce by fertilization. There are two kinds: Monocotyledons have a single seed leaf in the embryo, and dicotyledons have two seed leaves in the embryo.

Tomatoes are fruits that contain seeds. This characteristic places them in the angiosperm group, which reproduce by fertilization.

The Kingdom of Animals

Animals are multicellular organisms that are not plants or fungi. They are eukaryotes because their cells contain nuclei, and heterotrophs because they obtain energy by eating other organisms or organic matter. Most animals have true body tissue, collections of cells that perform certain body functions.

Animals can be divided into vertebrates and invertebrates. Vertebrates are composed of animals with spines, or backbones; invertebrates are animals without backbones. Invertebrates evolved first: The earliest animal fossils date back to about 600 million years ago, during the Precambrian era, and are those of invertebrates, including jellyfish or corals. Vertebrates, beginning with jawless fish, evolved about 500 million years ago. Land animals evolved about 400 million years ago.

INVERTEBRATES

Invertebrates can be broken down into categories based on their body tissues or body form. Parazoans are organisms that lack true tissue. Eumetazoans have true tissue, which develops after repeated cell division in the embryo. Invertebrates may also be distinguished by their symmetry, or balance of the organism's two sides. Types of symmetry include

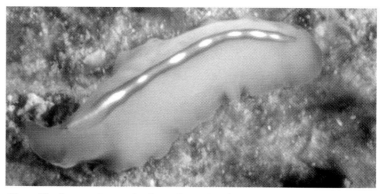

Above: The flatworm is a marine invertebrate classified as an acoelomate. It has neither a spine nor a central body cavity, but it does display bilateral, or mirror-image symmetry. Top left: The Christmas tree worm and the coral, photographed in Curaçao, are both invertebrates. The worm builds a tube on the surface of the coral; as the coral grows, the worm is protected within this tube.

bilateral (mirror-image) symmetry and radial symmetry. Animals with bilateral symmetry are called the bilateral or two-sided animals and include vertebrates and some invertebrates. Other invertebrates are radiolarians; they have portions of the body that radiate outward from a spokelike center.

Another way to classify invertebrates is through the presence of a body cavity. An acoelomate—a flatworm, for example—has no central body cavity; a coelomate is a bilateral invertebrate with a central body cavity, or coelom. Pseudocoelomates are organisms with false cavities and are considered to exist midway between coelomates and acoelomates. More developed than

acoelomates, coelomates are represented by annelids, such as segmented worms. Similar to them are mollusks, such as clams, octopuses, and oysters.

ARTHROPODS

Arthropods are invertebrates distinguished by having jointed legs. There are five classes of arthropods: crustaceans (such as lobsters), arachnids (such as spiders), chilopods (such as centipedes), diplopods (such as millipedes), and insects (such as butterflies). In all, there are between one and two million known species. Arthropods have bilateral symmetry, with bodies divided into jointed segments, an open circulatory system, and dependence on molting to allow

Above: The ascidian, or sea-squirt, is a type of chordate, an animal with a cord that stiffens the back. Top: The millipede, whose name means "a thousand legs," is an arthropod, an invertebrate that has jointed legs.

for growth. Each class of arthropod is characterized by distinctive behaviors and appearance: the arachnids by spinnerets, organs used for making webs; the millipedes by their many legs.

VERTEBRATES

The phylum Chordata includes all organisms possessing a notochord, a flexible, rodlike cord along the back. Vertebrates are part of this group, but, unlike other chordates, a vertebrate's notochord develops into the vertebral column. This characteristic, along with the braincase, or skull, and vertebrates' tendency toward cephalization, or concentration of neural and sensory organs in the head, distinguishes vertebrates from other chordates.

Vertebrates are a diverse group that includes fish, amphibians, reptiles, birds, and mammals. Fish, amphibians, and reptiles are alike in that they are cold-blooded, while birds and mammals are warm-blooded, but within each of those categories, organisms exhibit marked differences. For instance, fish live underwater and breathe through gills while amphibians live only part of their lives underwater and the rest on land, breathing through lungs. The remaining vertebrates are land dwellers all their lives. Reptiles are distinguished by dry, scaly skin; birds have feathers, and most can fly. Mammals nurse their young, and most have hair and reproduce through live birth in contrast to the other vertebrates, which reproduce by laying eggs.

Mosses and Evergreens

Plants are classified broadly as flowering or nonflowering varieties. There is great diversity in the nonflowering plant category. It includes towering specimens such as evergreen trees, which do not flower but nevertheless have leaves, stems, and roots. It also includes the low-growing mosses, which do not have true roots but still conduct photosynthesis like other plants.

The principal division in the realm of nonflowering plants is between bryophytes, nonvascular plants such as mosses, and tracheophytes, vascular plants such as evergreens.

BRYOPHYTES

A small group of nonflowering plants belongs to a division of the plant kingdom called Bryophyta. There are about 16,000 species of these bryophytes, or nonvascular plants. Bryophytes have no internal vessels or leaf veins to conduct water; nor do they have true roots. Generally they are small plants such as mosses, liverworts, and hornworts that thrive in moist conditions. Mosses (from the Old English word for "bog") are small brown or green plants with short stems and slender leaves that form a blanket on rocks, trees, or the ground. Liverworts are small

plants that absorb water over their entire surface. The liverwort is also called liverleaf. Hornworts are very small: They grow to only .375 to .75 inch (.95–1.9 cm) across.

Bryophytes absorb water and nutrients through hairy, root-like growths called rhizoids that anchor the plants to the soil. Leafy bryophytes require water for sexual, gamete-producing reproduction. The gametes disperse in water.

TRACHEOPHYTES

Most nonflowering plants belong to a division of the plant kingdom called Tracheophyta, or vascular

plants. This means that they have trache, which are vessels that run in a system through the entire plant. These vessels, xylem and phloem, conduct the transport of food, minerals, and water. They can be seen in the patterns of leaf veins in vascular plants.

Tracheophytes come in two principal subdivisions. One is Lycophytina, a group that includes club mosses and other plants whose leaves have a single central vein. These plants reproduce by spores that are produced in kidney-shaped sporangia in the stem-leaf nodes or on the leaves. These plants tend to grow in

Above left: The spore-containing sacs known as sporangia are clearly visible in this colored micrograph. Above right: The great horsetail is a nonflowering tracheophyte. Top left: The Ginkgo biloba, *which first evolved over 150 million years ago, is classified as a gymnosperm.*

Conifers have needlelike leaves that help them survive in cold, dry conditions. Less surface area leads to reduced moisture evaporation.

Evergreen trees, part of the conifer group, photographed in the Canadian Rockies. Conifers, named for their seed-bearing cones, are gymnosperms; like the ginkgo, they reproduce with bare seeds that are not enclosed within fruits.

moist, shady areas such as forest floors and near streams or rivers.

Euphyllophytina is the other main subdivision of tracheophytes. These plants usually have leaves with many veins. The nonflowering members of this subdivision include several classes: Psilopsida, or whisk ferns or fork ferns; Sphenopsida, or horsetails or scouring rushes; Pteridopsida, or ferns; and Gymnospermopsida, or gymnosperms. Gymnosperms reproduce by means of naked, or uncovered, seeds and include conifers, cycads, ginkgoes, and gnetaleans.

EVERGREENS AND FERNS

The primary group of seedless vascular plants are ferns. A fern is a thin plant with fronds, or leaves that grow outward from either side of a central stem. Ferns reproduce by way of spores. The fern dates back to the Carboniferous period, 354 to 290 million years ago.

Other vascular plants reproduce by means of seeds. They are gymnosperms, or vascular plants with bare seeds that are not enclosed within fruits. One type of gymnosperm is the conifer, or cone-bearer. Conifers produce cones that hold the otherwise unprotected seeds. Many conifers are cone-bearing evergreen trees. Their needlelike leaves are adaptations to cold, dry conditions: The needle shape allows the plant to reduce moisture evaporation, or drying. Conifers include pine trees, fir trees, spruce, larch, hemlock, and yew bushes.

Flowers and Fruits

The word "plant" literally means "a sprout," and flowering plants sprout fruits and flowers that provide practical reproductive use and aesthetic appeal. These flowering plants comprise Angiospermae, a class within the subdivision Euphyllophytina and division Tracheophyta.

Angiosperms are characterized as vascular plants with fruit-covered seeds or flowers, or both. They are the second major group of vascular plants (along with gymnosperms) and derive their name from their enclosed seed *sperm* holders *angi-*. These plants differ from gymnosperms, or seedless vascular plants, because their seeds are enclosed in fruits. They are not exposed, like those of the gymnosperm. Furthermore, their seeds are fertilized through a part of the plant.

GAINING DOMINANCE

From the Cenozoic era of 65 million years ago to today, angiosperms have dominated the plant kingdom. They are the most widespread and numerous of plants. Currently there are about 275,000 species of flowering plants. But when these plants first appeared, during the Mesozoic era (248 million to 65 million years ago), they were outnumbered by nonvascular plants and seedless vascular plants such as ferns and conifers. Over time, the reproductive capabilities of flowering plants prevailed: The flowering plants could be pollinated or be assisted in pollination by flying insects such as bees and butterflies and animals such as birds and bats. Fruits encased and protected their seeds, which aided in their propagation.

THE FLOWERING PLANT

The flower is the part of the plant involved in reproduction. It produces the seeds. A flower is located at the end of a stalk called a pedicle. Much of the plant, including the pedicle, is covered by a layer of protective cells called the epidermis. In the middle of the flower is the ovary,

Above: Tulips are one of more than 200,000 known species of flowering plants. These plants comprise the class of Angiospermae, *or vascular plants whose seeds are enclosed. Top left: A yellow bell pepper's enclosed seeds classify it as a fruit, though we commonly regard peppers as vegetables.*

The inside of a tulip, showing the stamens, or the male reproductive organs, grouped around the central carpel, the female part of the plant. The name carpel means "little fruit," as the seeds are contained inside. Flowers rely on cross pollination to reproduce.

To help with seed dispersal, burdock weeds have hooks attached to their seeds. The seeds are caught on animal fur and spread to new locations.

which produces eggs. Around the bottom of the flower are two or more sepals, outer coverings that protect the flower bud. Just above the sepals are the petals, which create a corolla, or crown, around the ovary. Inside the flower are the stamens, a group of vertical threads that grow above the ovary.

The male reproductive organ of the flower, the stamen consists of the filaments and anthers. In the middle of the flower is the carpel, which means "little fruit."

The fruit contains the seeds and is the female part of the plant. It consists of three parts—the stigma, style, and ovary. The ovule, which is the center of reproduction, lies within the ovary. The angiosperm shares similarities in its reproductive stages with the gymnosperm. The process is much like that of

the pinecone, which contains a conifer's seeds. The difference in reproduction for the angiosperm lies in its fruit, which is the mature ovary of a flower that has eggs or ovules covered by flesh.

Gymnosperms have naked seeds on the edges of their female cones. Angiosperms have seeds that come from their flowers and are then protected within its fruits. In some cases, the fruit can be eaten by humans and other animals.

Reproduction depends on cross-pollination. If a flower has not been pollinated, it will eventually wither and die. Most flowering plants reproduce by cross-pollination. That means that the plant does not pollinate its own flowers, but depends on insects, animals, and wind to deliver its pollen to another flower.

CARRYING THE SEEDS

Most seeds end up close to their original source. But plants can arise unexpectedly as animals inadvertently perform a service for them by carrying their seeds to new locations. Fruits evolved as a mechanism by which plants use birds and mammals to transport their seeds for them. Birds and mammals, attracted by the look and scent of fruit, eat the fruit, pass the seeds through the digestive system, and drop the seeds in their own fertilizer in new places. Flowering plants have different adaptations to procure transport for their seeds. For example, weeds called burdocks have hooks that catch onto animal fur; other fruits are similarly sticky, prickled, or barbed, and thus able to hitch a ride on animals.

Animals without Backbones

A nimals are divided between vertebrates (animals with backbones) and invertebrates (animals without backbones). Both vertebrates and invertebrates have the shared characteristics that define them as animals: They are living and breathing multicellular organisms that are not fungi or plants; they survive by eating organic material; and they are eukaryotes, composed of cells with nuclei. But the differences between them are marked.

EUMETAZOANS AND PARAZOANS

The older of the two types of animals are the invertebrates. There is great variety among them and they can be organized in many ways. One way of dividing them is as eumetazoans and parazoans. Most invertebrate species are eumetazoans, which are animals with tissue that forms after cell division in the embryo, and which have cells specified to perform certain body functions. For example, a lobster (a member of the phylum Arthropoda) has soft muscle tissue within a segmented, horny body covered with tough epithelial tissue.

Parazoans make up a small percentage of invertebrates and are notable for not having true tissue. The cells of sponges (members of the phylum Porifera), for example, have some specialization, enough to attach themselves to the ocean floor, but not quite enough to make individual tissues.

MANY PHYLA

Invertebrates come in many basic body plans, represented by the more than 30 phyla to which they belong. These are the some of the most important phyla:

- **Annelida** (segmented worms): These worms, such as earthworms and leeches, have bodies consisting of segments. Many worms have tentacles on their heads and a pair of leg-like projections called parapodia on each body segment.
- **Arthropoda** (arthropods): These animals have jointed legs, segmented bodies, and

an outside shell called an exoskeleton. They include insects, such as ants and beetles; crustaceans, such as crabs and lobsters; arachnids, such as mites and spiders; centipedes; and millipedes.
- **Brachiopoda** (lamp shells): These ocean-dwelling animals have a hard two-part shell covering a soft body.
- **Bryozoa** (bryozoans): These animals live in water, most

Above: Crustaceans like lobsters belong to the phylum Arthropoda, which have jointed legs, segmented bodies, and exoskeletons. Other arthropods include insects, such as ants and beetles.
Top right: Scallops are brachiopods, defined as having two hard shells enclose a soft body.
Top left: Jellyfish are members of the phylum Cnidaria, along with sea anemones and corals.

of them in colonies. Some colonies are jellylike masses, while others are networks on water plants.

- **Cnidaria** (cnidarians or coelenterates): These invertebrates are shaped like a cylinder, bell, or umbrella. Their bodies house a jellylike material between two layers of cells. Examples include jellyfish, sea anemones, and corals.

- **Echinodermata** (echinoderms): These spiny-skinned creatures have an internal bony skeleton. They have tubelike structures called tube feet. They include starfish, brittle stars, sand dollars, and sea urchins.

- **Mollusca** (mollusks): Mollusks make up the largest group of water animals, though some species live on land. Most mollusks have a hard shell that protects a soft body. The phylum includes clams, mussels, octopuses, oysters, snails, and squids.

- **Nematoda** (roundworms or nematodes): Many roundworms live in soil, water, or dead tissue. Some of these worms are parasites that are found in living plants and animals.

- **Platyhelminthes** (flatworms): Many flatworms live as parasites in other animals. Flatworms have soft, thin, flattened bodies with three layers of cells.

- **Porifera** (sponges): Sponges attach themselves to rocks and other objects at the bottoms of oceans, lakes, or rivers.

- **Rotifera** (rotifers): Rotifers live in lakes, rivers, streams, and oceans. They have cylinder- or vase-shaped bodies.

This sand dollar, washed ashore with beach pebbles, is related to sea stars and sea urchins, all members of the phylum Echinodermata, having spiny skins and internal bony skeletons.

The spider (above left) and the ant (above right) belong to the phylum Arthropoda. Spiders have two body segments and four sets of legs, while insects have three body segments and three sets of legs.

INSECTS VERSUS ARACHNIDS

Insects and arachnids are two of the five classes of the phylum Arthropoda, or arthropods, which means "jointed feet." From that shared classification, the differences begin. There are one to two million species of insects, but only 57,000 species of arachnids. In fact, there are more known species of insects than species of all other types of animals. (The Latin roots for insect represent "cut into": *in-*, for "into," and *sect*, for "cut," describing insects' segmented body structure. For arachnids, the Latin root is *arachn* for "spiders" or "webs." Common insects include flies and butterflies; arachnids include spiders and ticks.)

The body design of the two types of organisms also differs. The cut, or subdivided, insect has three body segments. In most cases, an insect is made up of a head, a thorax, and an abdomen, and usually three pairs of jointed legs and one or two pairs of wings. An arachnid has two segments: head and thorax, or cephalothorax. It has multiple (usually four) sets of legs for walking.

The Backboned Animals

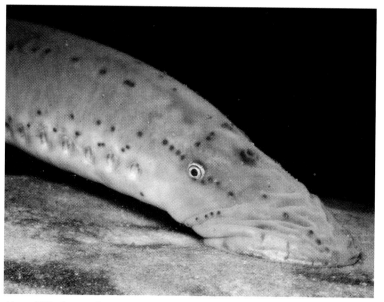

About 500 million years ago, long after multicellular invertebrates appeared, animals with backbones evolved. Known as vertebrates, they belong to the phylum Chordata, which is characterized by the presence, in at least part of the life cycle, of a chord, or cord, to stiffen the back.

In some members of the phylum Chordata, the notochord remains in the organism throughout its life; in others, the vertebrates, it is replaced by a full vertebral column, or jointed backbone. Two groups that retain a notochord throughout their lives are the urochordates, exemplified by the sea squirt, and the cephalochordates, such as the lancelet. While both types of organisms have notochords, they are officially invertebrates, although many scientists consider them important bridges between invertebrates and vertebrates.

Vertebrates (subphylum Vertebrata) are chordates with a jointed backbone and a braincase, or cranium. They have

Above: This chestnut lamprey, photographed in Arkansas, is a member of the class Agnatha, which comprises fishlike creatures without jaws. The class is among the oldest vertebrates on Earth. Left: Illustration of a vertebral column, common to all mammals. Top left: Lancelets are small, primitive chordates that live in temperate seas.

a high degree of cephalization, which means that the sensory, motor, and other nerve functions are concentrated in the head area, and the brain is in the skull. Though most have a notochord during the embryonic stage of their development, this flexible rod of cells is replaced by the vertebral column. The vertebral column houses a dorsal nerve cord, or spinal cord. The bones or cartilage form a lengthwise axis along the animal's back. Vertebrates have a closed circulatory system.

EIGHT TYPES OF VERTEBRATES

Vertebrates come in eight classes:
- **Class Agnatha** consists of fishlike creatures without jaws. These organisms were probably among the first vertebrates to make their appearance on Earth, about 500 million years ago. The class now includes about 60 species, mainly lampreys, eel-shaped vertebrates with gill slits.
- **Class Placodermi** consisted of jawed fishes called placoderms, which had flat surfaces and armor. They are now extinct.

- **Class Condrichthyes** is a group of fish with skeletons made of cartilage, such as sharks and sea rays. They have hinged jaws and paired fins; sharks have the added advantage of teeth.
- **Class Osteichthyes** consists of fish with bony skeletons. The 30,000 known species in this class include perch, salmon, and trout.
- **Class Amphibia** consists of four-footed animals that hatch from eggs and move from water to land. Members include frogs, newts, salamanders, and toads.
- **Class Reptilia** consists of organisms with waterproof scaly skin, including alligators, crocodiles, and turtles. They move by crawling, breathe through their lungs, and give birth from eggs.
- **Class Aves** is that of birds. They are warm-blooded and have wings, feathers, two legs, and a bill or beak, and they also lay eggs.

- **Class Mammalia** contains warm-blooded vertebrates that have mammary glands, which are used by mothers to produce milk for their nursing infants.

GROUPS OF MAMMALS

Among mammals are three major groups: monotremes, marsupials, and placentals. Monotremes have a single opening for their urinary and reproductive regions. Close to the reptiles in design, the class now includes mainly the platypus and spiny anteaters. Female marsupials have pouches in which offspring complete their development. Examples include the koala bear and kangaroo. Most mammals are placentals, which means they receive nourishment when young from a placen, or flat cake—the placenta. The fetus develops in the womb and receives nutrients from an umbilical cord attached to the placental organ.

Modern placental mammals include a number of orders, including Rodentia, toothy gnawers such as beavers, mice, and squirrels; Carnivora, carnivores such as lions, bears, and wolves; Artiodactyla, such as cattle, deer, and pigs, with an even number of toes on their limbs; and Chiroptera with 1,115 species of bats—the only flying mammals. Cetacea, large sea animals such as dolphins, porpoises, and whales; and Primates, such as apes and humans, with five-fingered hands and five-toed feet.

Right: Humans belong to the Mammalia, consisting of warm-blooded vertebrates with milk-producing mammary glands. Above: An egret belongs to the class Aves.

A Circle of Dependence

Seventeenth-century English poet John Donne spoke for humanity when he wrote, "No man is an island." Biologically, the interdependence extends further and is particularly noticeable among plants and animals.

Plants are able to make their own food, but they need the Sun's energy. Animals rely on plants' ability to feed themselves when they eat the plants for energy. Many animals also eat other animals for energy.

Plants and animals depend on each other to provide oxygen and carbon dioxide. Plants give off oxygen. Animals (including humans) breathe in the oxygen and give off carbon dioxide, which the plants require for photosynthesis.

THE PREHISTORIC PAST

The interdependence of plants and animals has evolved over millions of years and changed drastically many times. For example, about 65 million years ago, at the end of the Mesozoic era, a comet or asteroid struck Earth.

According to the most common view of the mass extinction that took place at this time, it was caused by a cloud of dust that cut off most of the sunlight, which killed plants. That left nothing for plant-eating dinosaurs to eat, and they died.

When plant-eating dinosaurs died, meat-eating dinosaurs had nothing to eat, and they died. This condition accounted for the extinction of the dinosaurs.

For every calamity like this, there were many instances of steady coevolution, in which plants and animals evolved in tandem to take advantage of each other's presence. As a result of this coevolution, plants assure wide dispersal of their seeds by having animals be their agents. For example, cherry trees have fruits that birds take in; they excrete the stones wherever they land, thus spreading the cherry tree seed. Plants such as grasses also benefit from herbivores that eat surrounding plants and allow the shorter grasses to receive sunlight.

THE HUMAN ROLE

Like other animals, humans are able to influence the ecosystem, but their domination of the planet allows them to do so with

Above: Cherry trees are part of the cycle of coevolution, by which plants and animals have evolved in mutual dependence. The tree's fruits are eaten by birds who disperse the seeds in their excrement. Top left: Agricultural fields are evidence of human dependence on plants for food.

exponentially greater power. In the Pleistocene era, several large animals were probably driven to extinction by human overhunting, a practice that has eradicated or diminished species in the past thousand years as well. Animals wiped out due to overhunting in the Pleistocene era include the giant kangaroo, the giant wombat, the giant armadillo, litopterns, and the giant ground sloth.

In the Middle Ages, in Europe, wolves were considered werewolves and believed to be carriers of disease. Efforts to rid the nations of wolves made the animal extinct in Europe. For different reasons, the same fate occurred for other animals in Europe, such as bear, bison, boar, deer, and elk. Today in Europe, most of these animals are found only in Russia and Scandinavia. In the United States, many fish and other mollusks such as oysters are currently facing shortages along the East Coast due to overharvesting.

Since humans began practicing agriculture, the Earth's ecosystems have suffered other disruptions. For example, the cultivation of a single crop may damage the soil. Yet despite these disruptions, there continue to be beneficial interactions between plants and animals, including humans. Bees pollinate flowers; flowers and plants give off oxygen; animals follow their natural cycles. Life continues.

The chemical predecessor to the analgesic aspirin is found in willow bark extract.

MEDICINAL PLANTS

Many plants have been and continue to be sources of human healing. The chemical predecessor to the analgesic aspirin was extracted from the bark of the willow tree. Ephedra, commonly known as horsetail plants, are a group of gymnosperms that contain the chemical ephedrine, which is used as a bronchodilator to ease breathing for asthma patients. It is also a blood vessel constrictor of linings of nasal passages; this reduces inflammation caused by hay fever. *Thymus vulgaris* is the source of thymol, which is used in a topical antifungal medication.

Accepted increasingly for traditional medicinal use are herbal remedies. Among the many in use are ginseng (*Panax ginseng*), a Chinese herb now used as tonic, stimulant, and mental revitalizer, and ginkgo (*ginkgo biloba*), an ancient tree used as a circulatory stimulant and tonic and as an anti-inflammatory.

Whales left on a beach after a hunt, around 1900. Commercial whaling has been globally banned since 1986. Whales are among the populations of wildlife threatened with extinction due to over-fishing, over-hunting, or over-harvesting.

CHAPTER 11

THE HUMAN ANIMAL

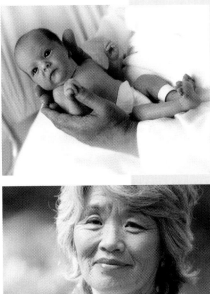

Left: Humans live in social groups, as do other animals. The traits that we share with other species are of as much interest to the biologist as those traits that make us unique in the animal kingdom. Top: A newborn baby, whose genetic makeup may determine much about its future development. Bottom: Scientists have made great advances in the study of human aging, but there is still much to be learned about nearly every facet of our biology.

For all their unique accomplishments—pyramids, opera, philosophy, airplanes, to name a few—humans are still animals. Close cousins of chimpanzees, they are members of the primate order in the class Mammalia and the phylum Chordata. A human is composed of more than 10 trillion cells, is physiologically active, interacts with its ecosystem, and lives in social groups analogous to those of other mammals. For all these reasons, humans are a fit subject for biological scrutiny. But there are things that make humans distinctive—including their language, their intelligence, and the fact that only humans can be biologists. The study of humans is of interest both because of what humans share with other animals and because of what humans alone possess.

Scientists have made remarkable advances in nearly every aspect of human biology—how embryos develop and adults age, the architecture of the brain and its relation to thinking and feeling, how individual behavior and social structures may be influenced by genetic imperatives, and how to treat diseases that plague us. But much remains mysterious about the biology of human beings. In some ways, the study of the human animal has only just begun.

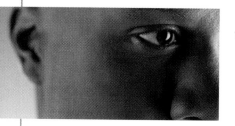

What Makes Us Human?

In terms of classification, structure, and function, humans are similar to other organisms. In their anatomy, humans are similar to other animals in that they have circulatory, respiratory, digestive, excretory, and reproductive systems. Like other vertebrates, they have a backbone, and, like other mammals, they nurse their young. Yet in certain respects they are very different from other animals.

BIPEDALISM AND BRAINPOWER

Many of these differences are tied to two features of humans: their two-footed, upright, striding walk (bipedalism) and a brain that is large in proportion to body size. The two-footed walk has allowed humans to evolve many adaptations, for example, a larynx that is situated in the body in such a way as to permit the development of language, and appendages (hands) that, because they are not needed for walking, have instead become refined for specialized use.

The proportionately large human brain has greater capacity than the brains of other animals, as is evident in the ability of humans to reason, conceptualize, and manipulate elements of the world, particularly

themselves. Because humans have larger brains than other animals do, they can reason as well as react instinctively. The building of a human society of law and government is the result of conscious and reasoned thought, as are the conception of beauty and the practice of art.

HUMANS AND OTHER ANIMALS

Humans share many physiological similarities with other animals. One similarity pertains

to obtaining nutrition. Unlike plants, animals such as humans are heterotrophic. In other words, they ingest other material as food, be it plant or animal. But unlike animals, humans are able to combine or process various foods using the chemistry of cooking, and they enjoy the food for its aesthetic appeal as well as its hunger-reducing qualities. In addition, vertebrates (including humans) and some invertebrates share a similar bilateral symmetry in body form; they also have a

Above: Bipedalism, or the two-footed walk unique to humans, is a trait that has led to many adaptations. No longer needed for walking, hands became specialized tools. Top left: The size of the human brain has allowed humans to develop the ability to reason.

Above: Humans and other animals use the environment to sustain themselves. The difference is that humans act deliberately, as evidenced by the rise of agriculture. Humans also have the capacity to reflect on the results of their actions and to alter them when necessary.

head at one end and an anus at the other. In the head (cephalic) portion of the body are sense organs, such as the eyes, the ears, and the brain.

Similarly, other vertebrates have skeletal, closed circulatory, and respiratory systems. Other vertebrates also have two layers of skin: an outer layer (epidermis), which contains pigments, pores, ducts, and dead cells, and an inner layer (dermis), which contains sweat glands, blood vessels, nerve endings, and the bases of hair and nails. But humans differ from many mammals—including their closest cousins, the great apes—in having relatively little hair on their skin.

Another similarity between humans and other organisms is that they can affect an ecosystem, using that environment to sustain themselves. The difference is that the nonhuman organism acts instinctively, while humans act deliberately. For example, when masses of locusts swarm and descend on a field or forest region, they consume virtually all the plants and leaves. When humans overharvest a region for crop production or overdevelop a region for housing, they also disrupt and sometimes destroy an ecosystem. But unlike nonhuman organisms, humans do have the capacity to reflect on their destructive actions and alter them.

Right: A male gorilla, one of the primates most closely related to humans. Instead of a head-to-toe layer of fur, humans have relatively little hair on their skin.

From Conception to Death

Like the life of all animals, human life has a distinct cycle. The cycle begins with the union of the sperm and egg in conception and ends in death, when the biological systems cease functioning. Between those points of development are infancy, childhood, adolescence, maturity, and senescence.

BIRTH

Following the sperm's fertilization of the egg, the zygote develops into a cluster of cells called a morula. It enters the hollow opening of the uterus, becomes a blastula, and attaches itself to the interior lining of the uterus (the endometrium). From there, the cells become specialized and a gastrula ("little stomach") is formed. In the gastrula are layers of cells called derms. They include the endoderm (inner skin), the mesoderm, and the ectoderm. During the first three months of development, the organism is called an embryo. From then until birth it is called a fetus.

Nourished by the umbilical cord, which stretches from the embryo's midsection to the placenta, the fetus grows and develops its vital systems. After about 250 days, the fetus is no longer able to live within the mother's uterus due to its size and to the disintegration of the protective area surrounding it. Birth occurs when dilation increases the size of the cervical opening and the fetus passes down the birth canal into the outside world.

INFANCY THROUGH ADOLESCENCE

Infancy spans the first eighteen months of life after birth. During this time, an infant is able to gain some large motor control and typically attains developmental benchmarks that include the abilities to stand upright, walk, and talk. Sensory acuity develops rapidly. Infants are born with outsized heads to accommodate their large brains, but as they grow their body size increases, becoming more in proportion to the size of the head.

During childhood and adolescence, the skeleton grows in size and strength. A child's weight increases substantially up to the age of three; it will not increase as markedly until adolescence. During these early years, a child also develops spatial perception, which promotes hand-eye coordination, useful in everyday skills such as sports. Language

Above: The cycle of human life has distinct stages. Conception (top left) occurs when sperm fertilizes the egg. In females (left), puberty typically begins at between ages 8 and 12, spurred by the hypothalamus gland. Senescence, or old age, is the final stage of life, following maturity.

and socialization skills develop.

In puberty, primary and secondary sexual characteristics develop, adult reproductive capacity appears, and sexual interest surges. Puberty typically begins in girls between 8 and 12 years of age, whereas boys start about two years later. The hypothalamus initiates pubertal changes by directing pituitary growth hormones and the hormones that control the ovaries and testes. A girl's breasts grow, her pubic hair develops, and her body takes on rounded contours. The first menstrual period occurs at about age 12 or 13. A boy's genitals grow, his pubic hair develops, his voice deepens, and facial and underarm hair appear. The first ejaculation typically occurs sometime between the ages of 11 and 15.

ADULTHOOD THROUGH OLD AGE

In the adult, organs continue to function if injured, and some, such as the liver, are still capable of growth. But beginning in early adulthood, bone tissue breakdown slowly begins to exceed deposits. As a person ages, bone tissue is depleted, and the gradually weakening bones are increasingly susceptible to breaks. In women, there is a gradual decline in the function of the ovaries and in the production of estrogen. The average age at which menopause (the end of the menstrual cycle) occurs is about 50. In men, testosterone production declines over the years, and the testes become smaller.

As the body undergoes aging (senescence), hearing and vision decline, muscle strength lessens, soft tissues such as skin and blood vessels become less flexible, and body tone declines. Most of the body's organs perform less efficiently; for example, the heart pumps less blood per minute and the rate of antibody production drops. The aging brain undergoes a progressive loss of specialized nerve cells called neurons, although these losses represent only a small percentage of the total neurons in the brain.

Over time, the human body's reproductive systems slow. In women, the function of the ovaries and estrogen production begins to decline, with menopause occurring at about age 50.

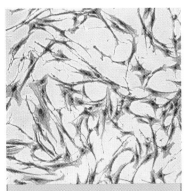

Aging cells continue to divide until they die, as shown in this light micrograph.

AGING CELLS

Every day, several billion cells in the human body die and are replaced by new cells. As a cell in the human body ages and declines, it acts to preserve the cells that surround it by undergoing autolysis, or self-breakdown. In this process, rupture of lysosomes (sacs in the animal cell cytoplasm) can occur. The rupture liberates digestive molecules that destroy the entire cell; the destruction minimizes the accumulation of dead cells and ensures that the unhealthy cell is kept away from adjacent healthy cells. The remnants of the dead cells pass out of the body with other waste products.

An Unusual Body Plan

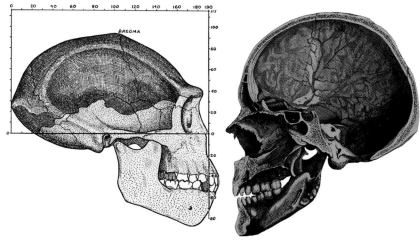

Above: A comparison between the skull of a direct human ancestor, Homo erectus, *and a modern human skull, or* Homo sapiens. *Over the past two million years, the average human brain has nearly doubled in size. Top left: Unlike similar primates such as the chimpanzee, humans are bipeds, meaning they have an erect posture and walk on two feet.*

With their distinctively shaped spinal column, two legs, large brain, and five-digit extremities with manual dexterity, humans have an unusual body plan. Unlike other primates, they also have particular strengths.

TWO-FOOTED WALK

Humans are distinguished by their skeletal structure and the adaptations that accompany it. Unlike similar primates such as the chimpanzee and gorilla, humans are bipeds, meaning they have an erect posture and walk on two feet. Unlike other mammals, they have an S-shaped spinal column, which puts the body's center of gravity above the feet for balance when the human is standing upright. The exact centers of gravity differ for males and females due to anatomical differences; the female center of gravity lies in the lower trunk, for childbearing purposes, while the male's center of gravity is in the upper trunk.

Humans also have other adaptations that accompany bipedalism. Among them are a broad pelvis, or hip, for stabilization (with the female's hips proportionately wider, for childbearing), a locking knee joint for posture and mobility, and a longer heel bone and longer and aligned big toe for stability.

Because humans do not have to use their hands to bear part of their body weight while walking as do other anthropoids, they have greatly developed the use of their hands. Their manual dexterity allows humans to find food and perform a variety of tasks. Of essential use for manipulating and grasping is the elongated human thumb, which is opposable to the other digits and can rotate.

OTHER REFINEMENTS

Erect walking also makes possible other uniquely human physiological refinements, such as speech. The physiological requirements for speech were secondarily established by erect posture, which positions the vocal cords for production of sounds. The control of the lips and tongue required for speech is also linked to erect posture. The skilled use of the hands, which is an adaptation of erect posture, occurs in association with the enlargement and specialization of the brain, which in turn is a prerequisite for refined control of the lips and tongue.

Humans have a proportionately large brain. Over the past two million years the average human brain has grown in size from about 44 cubic inches (721 cm³) to 85.4 cubic inches (1,399.5 cm³). This growth was made possible by neoteny, the retention of juvenile characteristics until late in life. In humans, childhood, with its rapid brain growth, is extended much later in life than it was in their ape ancestors. Indeed, the infant is born with only 25 percent of mature brain capacity. Neurological pathways are still developing and are influenced by external stimuli during childhood.

LEARNED BEHAVIORS AND CULTURE

These physiological adaptations permit a variety of human behaviors. Humans can adapt to environmental changes and act through learning rather than from instinct alone. Thus, humans are able to adapt to a wide range of habitats, rather than depending completely on natural selection to drive adaptation. Another important trait of humans is that they are omnivores (creatures that can derive nutrition from both animals and plants). Unlike mammalian carnivores or herbivores, the omnivore human is more adaptable, allowing it to sustain life in many different kinds of environments.

Consciously thinking and reacting, humans are able to engage in culture. They are able to develop new skills and teach them, build new kinds of social networks, and deliberately change their environment.

Above: The ability to consciously think, reason, and react allows humans to learn new skills, to teach these skills, and to deliberately affect their environment. In contrast, other species depend soley on natural selection to drive adaptation. Top: The omnivorous diet of humans, who derive their nutrition from both plants and animals, allows them to be more adaptable and suited to living in many different environments.

How a Human Works: Human Physiology

Humans operate through a set of bodily systems that are similar in many ways to those of other animals, yet nonetheless remarkable for their complexity. The most long-lasting components (aside from the teeth) are the 200 bones that comprise the skeleton. This hard, dried collection of bones that protects the internal organs is known as the endoskeleton. Held together by tough ligaments, the endoskeleton is part of the musculoskeletal system. Tendons attach skeletal muscles to the bone; these muscles contract to make skeletal bones move. Control over the body's muscles is exercised by the nervous system. The somatic portion of the nervous system permits voluntary control over skeletal muscles, while the autonomic, or involuntary, nervous system regulates cardiac and smooth muscle and glands. Nerve impulses in the

Left: The human skeleton is made up of 200 bones, held together by tough ligaments. Top: The somatic part of the nervous system controls skeletal muscles.

Though humans have three-dimensional color vision, which many other animals lack, they have a poorly developed sense of smell relative to animals such as dogs. Still, the human olfactory system is able to distinguish thousands of different smells.

part of the brain called the cortex cause the voluntary movement of the head, body, and extremities. While many body functions are controlled by the nervous system, some are governed by the endocrine glands.

The brain is informed about the environment through the senses, which in humans differ somewhat from those in other animals. Humans have three-dimensional color vision, which many animals do not, but their sense of smell is poor relative to animals such as dogs. Even so, humans are able to distinguish thousands of different smells.

CIRCULATORY AND IMMUNE SYSTEMS

The human body houses the many veins, arteries, arterioles, and capillaries through which the blood circulates, pumped by a four-chambered heart that is similar to the hearts of other mammals. The blood vessels of the human body form a branching network that is about 60,000 miles (96,558 km) long. Valves in the heart make sure the blood pumps in the right direction. By the time a person is 70, his or her heart will have pumped about 55 million gallons (208,197,649 liters) of blood.

Right: The airway system of human lungs are visible in this three-dimensional CT scan. Respiration takes place through the expansion and contraction of the lungs.

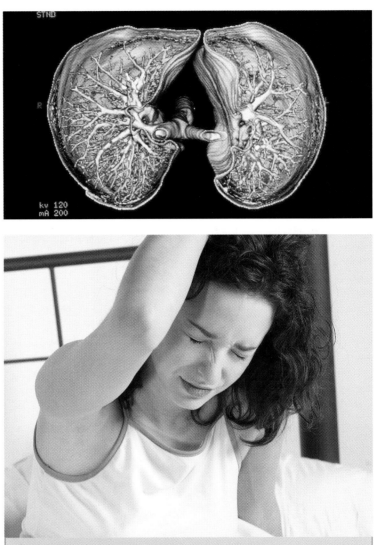

To defend itself against infectious microorganisms, the body has an immune system, which includes white blood cells, such as lymphocytes, manufactured in the bone marrow.

RESPIRATION, DIGESTION, AND EXCRETION

Respiration (the intake of oxygen and release of carbon dioxide) takes place through the expansion (inhalation) and contraction (exhalation) of the lungs. The pace of respiration and exhalation is regulated by the brain, and contractions of the diaphragm control changes in chest capacity.

The digestive system is activated by food. As food is chewed and mixed with saliva, it passes into the stomach. There, gastric and intestinal juices continue the process, resulting in a mixture called chyme. The chyme is pushed down the alimentary canal by peristalsis—the rhythmic contractions of the smooth muscle of the gastrointestinal system. Nutrients are absorbed into the body mainly through the small intestine. Waste substances from the liver and unabsorbed food pass to the large intestines and are excreted as feces.

Water and water-soluble substances move by the bloodstream to the kidneys, which act as a filter. The kidneys return most of the water and salts to the body; excess water, salts, and waste are excreted as urine.

Sleep deprivation is a common problem of modern life, causing serious health side effects. It is estimated that over 70 million Americans are sleep-deprived.

THE SLEEP-DEPRIVED ANIMAL

Culture and society sometimes get in the way of human physiological needs. That is the case with sleep deprivation, an increasing problem in the fast-paced modern world. According to the U.S. National Institutes of Health, more than 70 million Americans are sleep deprived. On average, they get two hours less sleep per night than their ancestors did a century ago. Chronic sleep deprivation can sicken people, leaving them more prone to disease, making them more dangerous drivers, and shortening their life spans—just three of the possible consequences. Evidence suggests that the immune system does not function as well in sleep-deprived individuals, leaving them susceptible to infections.

How Humans Behave

Human behavior can be socially neutral, destructive, or constructive. Humans have some behaviors in common with other species of mammals, while others are unique to humans. Some, such as seeking warmth, are instinctive; others, such as marriage, are tied to culture. But all of them together help to define us as human.

INSTINCTIVE BEHAVIOR

Instinctive behavior is made in response to physical stimuli and is not determined by culture. It includes such actions as scratching oneself, sneezing, shivering, squinting to shield one's eyes, seeking warmth or coolness in extreme weather, or cleansing oneself when soiled. Such behavior is considered neutral, without cultural significance. It is performed by both humans and animals.

CULTURE AND BEHAVIOR

Humans also engage in various types of culturally benign social rituals to mark personally important occurrences, such as birth, mating, and death. The birth of a new member of a group is marked by notification of the group and the government or group of ruling elders. Mating involves private and public behaviors. In most human societies, the choosing of a sexual mate involves some form of courtship. It may be some form of vocal interaction, mutual grooming, or other nonsexual physical encounter. Traditionally, marriage is a societal and governmental sanction of a male-female union, although some societies sanction same-sex and multiple-partner marriages.

Rituals and observances involving death and disposal of the human body are both universal and culture-specific to humans. Not only are they associated with religious beliefs about the nature of death and the existence of an afterlife, but they also have important psychological, sociological, and symbolic functions for the survivors. Thus, the study of the ways in which the dead are treated in different cultures leads to a better understanding of the many diverse views about death and dying, as well as of human nature. Funerary rites and customs are concerned not only with the preparation and disposal of the body but also with the well-being of the survivors and with the persistence of the spirit or memory of the deceased.

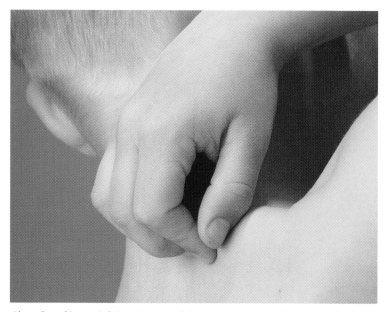

Above: Scratching an itch is an instinctive behavior, practiced by both humans and animals. Top left: In many cultures, marriage is a traditional social sanction of the choice of a mate of the opposite sex, though some cultures sanction same-sex or multiple-partner unions.

POSITIVE AND NEGATIVE BEHAVIORS

Some behaviors, such as the altruistic donation of services or goods, are almost inevitably praised by most societies. Saving a life is almost universally considered a behavior beneficial to society.

Other behaviors, however, are criticized and often regarded as negative. Blameworthy acts such as killing and torture are violent, yet violence in and of itself does not always receive social disapproval. The police officer or soldier killing in performance of duty may be praised as a hero, while the murderer killing for selfish, personal gain or out of a state of depravity is not.

Similarly, transforming a landscape or hunting a species to extinction may be considered a laudable act of pioneering or a reprehensible act of ecological destruction. What is certain is that humans, by virtue of their intelligence, have a great capacity to destroy as well as create. How society judges the exercise of that capacity varies from one context to another.

Left: Firefighters often endanger their own lives in order to save others. Saving lives, like other altruistic behaviors, is almost universally considered beneficial to society.

Left: A woman reads with a child. The perception of differences between adults and children has been observed as a universal of behavior, common to all societies.

EVERYBODY DOES IT

Some elements of human behavior are special to one society or even an individual. But many elements are common to all societies. This suggests that they are rooted in human genetic inheritance. These are some universals of behavior and overt language observed in all cultures, as compiled by anthropologist Donald E. Brown:

- Belief in the supernatural/religion
- Childhood fears
- Cooking
- Distinguishing right and wrong
- Government
- Hairstyles
- Language employed to misinform or mislead
- Law (rights and obligations)
- Male and female and adult and child seen as having different natures
- Narrative (storytelling)
- Promise
- Rhythm
- Sexual attraction
- Taboos
- Weather control (attempts to)

Plagues Abounding

L ike all other animals, humans are plagued by disease. Disease afflicts, deforms, disables, and kills people daily. Unlike other animals, humans are able to combat disease with the science and art of medicine, but our basic biological vulnerability to disease remains.

Diseases may be present in the body before birth, in one's genetic makeup (for example, Tay-Sachs and cystic fibrosis). They may develop over time, without outside causes, often through cell mutation (type 1, or insulin-dependent, diabetes is one example). They may also

Medical researchers cut open rats suspected of carrying bubonic plague in 1914. The disease was contained by rat-control measures.

develop over time with the aid of outside causes (lung cancer, for example). Finally, they may be transmitted by a carrier of the disease who passes it through bodily fluids released in the air (tuberculosis, for example) or through sexual encounter (AIDS, for example). Three common disease sources are bacteria, viruses, and cell mutations.

BACTERIA

Direct contact with a flea or with droplets of saliva coughed up by infected humans transmitted a bacterium called *Yersinia pestis,* which was responsible for the deaths of 25 million Europeans in the fourteenth century in a plague known as the Black Death. Septicemic and bubonic plague were transmitted by a bite from the oriental rat flea, and the pneumonic plague was carried through airborne droplets.

In the nineteenth and twentieth centuries, *Mycobacterium tuberculosis* caused many deaths, particularly in newly industrialized and crowded cities. The infectious disease of tuberculosis was and is spread through the airborne droplets conveyed by coughing or sneezing. As multiple people have contact with the infected individual, the disease spreads.

Above: Infectious diseases can be spread through airborne droplets emitted by sneezing. Top left: Some diseases are aided by outside causes, such as lung cancer from smoking.

VIRUSES

Viruses are responsible for many highly contagious and potentially fatal diseases.

One example is polio, also called poliomyelitis, which affected many thousands of children and adults in the United States and developed countries during the twentieth century. This highly contagious infectious disease is spread through a virus that lives in the throat and intestinal tract. Although most of the developed world has been vaccinated against the disease, it can be spread to those who have not been immunized through direct contact with infected stool or secretions from the throat. One variety of polio can cause paralysis.

Even more deadly was the Spanish influenza pandemic of 1918. Killing between 20 million and 40 million people in Europe, the United States, and worldwide, it is considered the most deathly epidemic in world history. In 2005, it was determined that the virus that is responsible for avian flu, which is beginning to cross the species barrier to infect humans, has a genetic composition that is similar to the virus responsible for the 1918 epidemic. Scientists are concerned that the avian flu virus will produce a pandemic (a global illness outbreak).

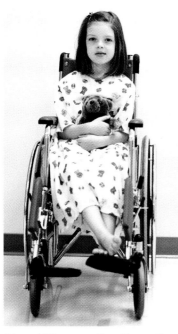

Many children were affected by a polio epidemic during the first half of the twentieth century. One variety of the highly contagious disease can cause paralysis.

MUTATIONS

A prevalent disease caused by cell mutation is cancer. It occurs when there is an abnormal change in the genetic makeup of the body's cells. Rather than duplicating and dividing to create normal daughter cells, the mutation creates cancerous daughter cells that multiply quickly and haphazardly and can travel throughout the body. The regulatory machinery that controls the cell division process is aberrant in cancer cells. As a result, the cells' growth and division continues, forming tumors that interfere with normal cell activity. Death may result.

Mutations can be triggered by toxic substances called carcinogens; chemicals in cigarette smoke are carcinogenic, for example. Thus, smoking is now recognized as a leading factor in lung cancer. The risk of skin cancer is aggravated by exposure to sunlight, as through sunbathing.

Sickle-cell anemia is a mutation that affects the development of red blood cells in some people of African descent. The sickle cell evolved as a protection against malaria.

THE VALUE OF SICKLE CELLS

The deadly diseases of malaria and sickle-cell anemia are prevalent in much of Africa, particularly in central Africa. Malaria arose first, and the sickle cell evolved as protection from it. It has been found that individuals who carry one copy of the sickle-cell gene, a mutation to the hemoglobin gene that affects the development of red blood cells, are less susceptible to the most common form of malaria than people without the gene. Those who inherit two copies of the sickle-cell gene usually develop sickle-cell anemia, a disease known widely in countries with descendants from Africa, including the United States. In these countries, the risk of malaria is often negligible, making the question of protection from malaria a nonissue. But in parts of Africa, where malaria is still present, the sickle-cell gene still provides some protection against malaria.

Are We Our Genes?

The recent success of the Human Genome Project in sequencing all the genes in the human genetic inheritance left open an important question. How much of what we are is written in our genes, and how much of it is written elsewhere—in the natural environment, society, or our own choices? What is the role of genes and environment in determining human destiny? Genomic research is helping to answer the question, but much remains to be learned.

NATURE VERSUS NURTURE

The issue of genes versus environment is part of an older controversy: whether nature or nurture has a greater role in determining human characteristics. Is the human simply a blank slate at birth waiting to be written on, as seventeenth-century English philosopher John Locke believed, or is much of who we are determined by heredity?

The confusing answer is both and neither. Genes do not contain a trait as an eggshell contains an egg. Rather, a gene contains instructions for making a protein. That protein, once incorporated into the body, contributes to the making of a trait, from physical ones such as height and skin color to behavioral ones such as intelligence and sense of humor. The contribution of any one gene may be large. For example, in the case of a genetically determined disease such as Huntington's disease, having the identified gene makes it virtually certain that one has the disease. But in many cases, including diseases such as breast cancer, a number of genes coupled with certain environmental triggers combine to produce a trait, so that the contribution of any one gene is small. Human characteristics are determined by both genes and environment, in differing proportions depending on the trait.

PSYCHOLOGICAL AND CULTURAL TRAITS

With psychological traits, a complex of genes and environmental factors is usually at play, making it difficult to tease apart the causes of those traits. Some insight can come through studies of identical twins reared apart, with different environments but identical genomes. The studies show that the effect of differences in genes on mental factors, such as intelligence, personality, and tastes, is substantially more than zero but considerably less than 100 percent.

In virtually all the traits involved in cultural behavior, many genes interact and environment plays a large role. For example, environment determines whether a child speaks Turkish or Japanese as a first language: Children can learn either language with equal ease, and the main issue is whether the child is being

Above: Studies of identical twins reared apart have shown that genetic factors contribute to their similar development and personality. Top left: In genomic research, questions are still unanswered. For example, how much of what we become is predetermined by our genetic inheritance?

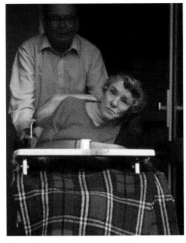

Some diseases are genetically determined, such as Huntington's disease; if one has the gene, it's almost certain the disease will develop.

raised in a Turkish- or Japanese-speaking household or community. Nevertheless, culture is not all-important. Many linguists believe that the deep structure of all languages is determined by universal rules that are innate in the human mind. Individual verbs and nouns are a product of cultural history, but all languages have verbs and nouns.

Similarly, specifics of the world's religions are not determined by the genetics of the faithful. But genes may contribute to the likelihood that a person will believe in God. The gene VMAT2, which influences the flow of mood-altering chemicals in the brain and which has been dubbed the "god gene," has been found to be linked to the ability to believe in a greater spiritual force. Even so, much remains unclear as to how genes interact with environment to produce the full range of human behavior and physiology.

Genes may contribute to belief in a higher power. Dubbed the "god gene," VMAT2 influences the flow of mood-altering chemicals and has been linked to a tendency toward spirituality.

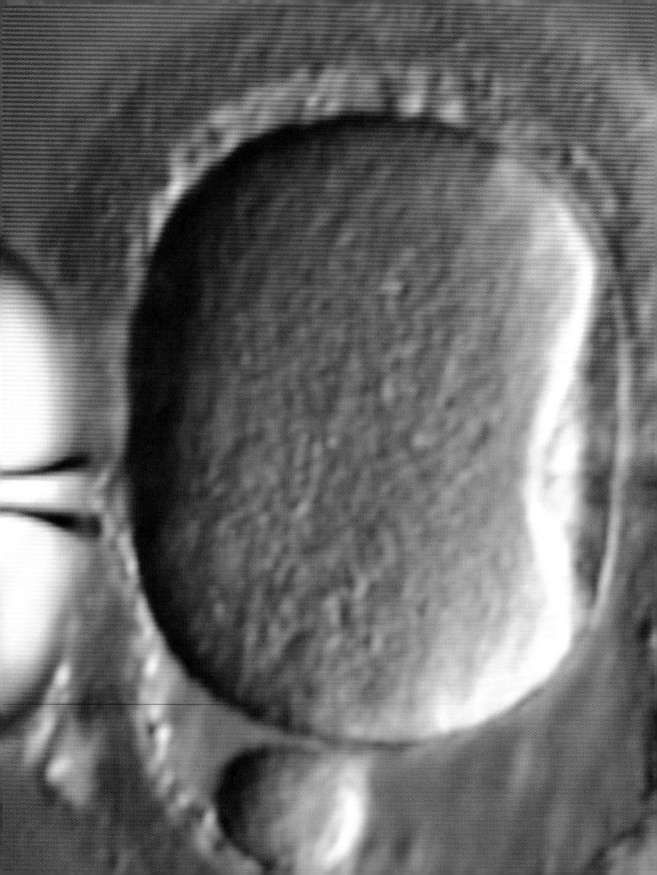

CHAPTER 12

A BREAKING SCIENCE

Left: In-vitro fertilization, or medically assisted procreation. The sperm is introduced to the egg outside a woman's body; the fertilized egg is then transferred into her uterus. Such fertility treatments are becoming increasingly common. Top: Medical researchers are working on new treatments for diseases like cancer and multiple sclerosis. Bottom: Several species of rain forest frog are facing extinction, a danger that has been linked to the effects of global climate change.

On any given day, some of the world's breaking news is related to biology. There might be controversies over the teaching of evolution in schools, or disputes about whether embryonic stem cell research is moral. Medical researchers might unveil a new cancer treatment, while others shows that a common personality trait is linked to a specific gene. Ecologists connect a decline in frog numbers to global warming. Animal researchers report on wolf behavior. Geneticists clone creatures, genetic engineers modify genes, and botanists discover a new kind of tree.

Biology today is fast moving and multifaceted, with much relevance for nearly every area of society. Biology is naturally of interest to those who like knowledge for its own sake, but many people follow the dispatches from this breaking science because they know it concerns them and their everyday lives.

The wave of news is likely to continue. Scientists have only begun to decipher the data from the Human Genome Project. Exploration of the brain really has just begun. The deep sea remains mysterious, and the issues at stake in developing "designer babies" have yet to be fully faced. Biology will be making news for a long time.

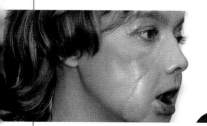

Medical Miracles

In November 2005, the first face transplant in history made headlines around the world. Frenchwoman Isabelle Dinoire, whose face had been mauled by her own dog, received a flap of facial tissue containing the nose, lips, and chin of a dead woman.

Controversy arose over whether the surgical team, led by Bernard Devauchelle, should have performed the risky surgery for a woman whose injuries were not life threatening. But no one doubted that the feat was impressive. The surgeons had to apply a tremendous amount of biological knowledge and medical skill to connect blood vessels and prevent the patient's immune system from rejecting the graft. Once again, biology had paved the way for a medical advance.

SURGERY

At the dawn of the twenty-first century, the medical frontier appears full of promise. Tissue transplants are only one option for patients who need replacement parts. Many medical researchers are developing bionic parts, electronic or mechanical components to replace damaged tissues or organs. A company called Optobionics has carried out a clinical trial in which surgeons implanted a .08-inch (2 mm) microchip in the eyes of

TOWARD A NEW FACE

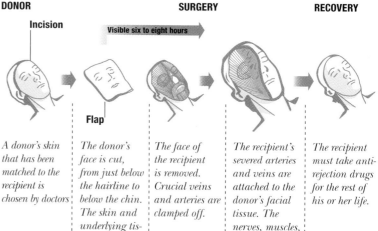

DONOR   **SURGERY**   **RECOVERY**

Incision

Visible six to eight hours

Flap

A donor's skin that has been matched to the recipient is chosen by doctors.

The donor's face is cut, from just below the hairline to below the chin. The skin and underlying tissue is removed.

The face of the recipient is removed. Crucial veins and arteries are clamped off.

The recipient's severed arteries and veins are attached to the donor's facial tissue. The nerves, muscles, and skin are connected.

The recipient must take anti-rejection drugs for the rest of his or her life.

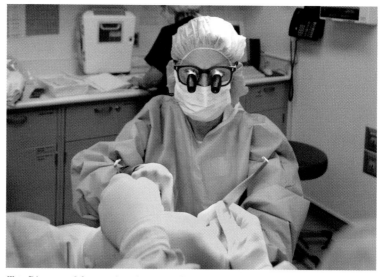

Top: Diagram of the steps in a face transplant surgery. The controversial procedure raises many moral concerns. Bottom: Microsurgery on a patient's face at the Cleveland Clinic in Ohio. Top left: A French woman, Isabelle Dinoire, received the world's first face transplant in 2005.

patients with retinitis pigmentosa, a degenerative eye disease. The result was impressive restoration of vision.

Surgery itself is being transformed by the introduction of robots that can allow surgeons to perform surgery on patients in distant locations. The surgeon views the patient on a video monitor while robotic hands, controlled remotely by the surgeon, perform the operation. In 2001, for example, a surgeon in New York used a computer and robotic tools to remove the diseased gallbladder of a patient in France.

CANCER

Another major area of medical progress in the early twenty-first century is oncology, or cancer research. Cancer researchers have found new ways of attacking tumors. One way is by fighting the growth of new blood vessels (angiogenesis) in the cancerous tissue. This can be done by using molecules that block angaiogenesis by acting as an "off" switch that turns off blood-vessel growth. The outlook for antiangiogenic therapy remains promising. By the early 2000s, at least 10,000 cancer patients had been treated with some variety of experimental antiangiogenic therapy. In 2003, large-scale clinical trials of the antiangiogenic drug bevacizumab (Avastin) demonstrates extended period of survival of treated cancer patients.

As oncological research has advanced, cancer drugs have been targeted to destroy only cancer rather than harming healthy body tissues along with cancerous ones—a problem that had plagued past treatments. In 2004, the U.S. Food and Drug Administration (FDA) approved two of these targeted pharmaceuticals for the treatment of colon cancer. "This is the most exciting time that I have known in the field of medical oncology," commented American Cancer Society president Ralph Vance.

MICROBIAL DISEASES

Even as some researchers combat cancer, others struggle with diseases caused by microbes. Avian influenza A, or H5N1, a virulent new strain of avian, or bird, flu that surfaced in Asia, poses a tremendous challenge to scientists. Although thus far it primarily has affected only birds, this deadly new strain holds the devastating potential to infect people. The great fear is that it will mutate into a form that could be transmitted from person to person, causing a global pandemic that would kill millions, like the influenza pandemic of 1918.

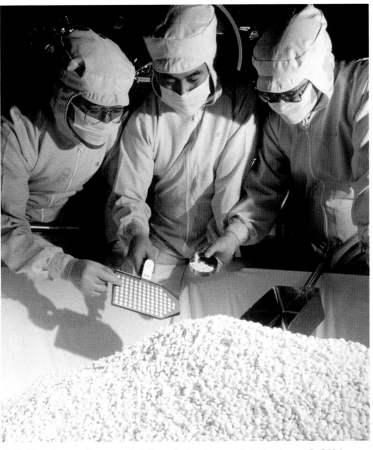

Technicians inspect pharmaceutical drugs during the manufacturing process. In 2004, new drugs that targeted only cancerous body tissues were approved for the treatment of colon cancer. Medical scientists have made great strides in the advancement of oncological research.

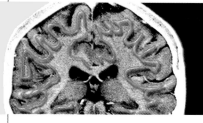

The Mind and the Brain

Agrowing area of biological research is that of mind and brain—the way that people think and feel (mind) and how that activity is produced on the levels of cells and tissues (brain). The years 1990–2000, declared the Decade of the Brain by U.S. presidential proclamation, produced numerous innovations and discoveries. These included the growing ability to use scanning technology to monitor changes in blood flow in the brain. Known as functional imaging, this technique showed which parts of the brain were active during various tasks, providing a window into how the brain performed its functions. Discoveries into the mind and brain continue into the twenty-first century.

MAPPING THE BRAIN

One aspect of mental life that has long been mysterious is volition, or the will. Increasingly, a link has been found between volition and a particular part of the brain called the anterior cingulate cortex (ACC). In 2002, one study showed that early activation of the ACC during a task in which volunteers had to make a choice reflected their intentional effort to carry out the task.

Other parts of the brain handle other tasks, sometimes with a high degree of specialization. In 2004, neuroscientists found that the brain has a particular area whose purpose is to recognize faces, and other neural pathways for recognizing other sorts of things, such as houses.

HELPING THE SICK

Mind and brain research has practical applications for medicine. It was long thought that damaged cells in the central nervous system could not regrow, but current research is showing otherwise. In October 2005, American neurologist Wendy Kartje reported that damaged nerve cells in rats could grow new neural connections, restoring lost leg function. She achieved this result by preventing the natural brain protein Nogo-A from binding to nerve cells. The technique might one day help human stroke victims regain lost speech and movement, for example.

Another promising procedure for people paralyzed by neurological injury comes from technology that directly links the brain to a computer. In

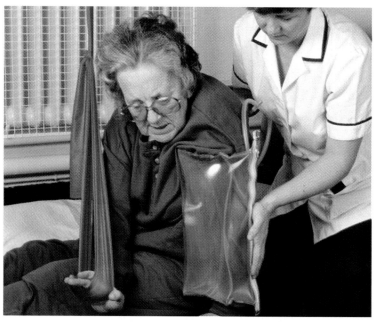

Above: Physicians caring for stroke victims can now hope for regained speech and movement. Top left: New techniques for mapping the brain, such as this Magnetic Resonance Imaging (MRI) scan of a normal human brain, help to monitor blood flow in the brain.

2004, young American Matthew Nagle, who is paralyzed in all four limbs, was outfitted with a brain implant that allows him to control a computer and devices connected to it, such as a television. He has also learned to open and close an artificial prosthetic hand through the brain implant. Much more progress is needed, but in time the technology may help other people with severe paralysis.

THE SCIENCE OF HAPPINESS

Even as neuroscientists strive to understand the workings of the physical brain, psychologists try to understand the functioning of the mind. One area of research has been happiness, or subjective well-being, a person's own judgment of the state of his or her life. Psychologist David Lykken has described various actions that people can take to increase their levels of happiness, including focusing on things that give them pleasure and counteracting negative emotions. Studies reported by Daniel Goleman provide scientific evidence that compassion, as stimulated by Buddhist meditation techniques, can cause a state of joy. This finding is one sign of how modern biology can sometimes reach back to connect with ancient sources of wisdom.

Right top: Neuroscientists and psychologists are working to understand brain activities that influence a person's sense of well-being. Right botttom: A Tibetan Buddhist monk. Studies provide scientific evidence that compassion, like the sort practiced in Buddhist meditation techniques, can induce a state of joy.

Genomes Unveiled

On April 14, 2003, the Human Genome Project announced completion of its multiyear initiative to sequence all three billion DNA letters in the human genome. The result is a tremendous database of genomic knowledge that is being analyzed in efforts to understand how genes cause disease and contribute to physical and personal human traits. Meanwhile, projects to sequence the complete genomes of other organisms have also been completed. The translation of all that data into practical applications has only just begun.

THE GENOME AND MEDICINE

One application of genomics is to understand which specific genes are involved in human diseases, in the hope of treating those diseases. For example, in 2003, scientists published evidence that bipolar disorder—a mental illness in which patients suffer mood swings from mania to depression—is linked to a gene on chromosome 22. The gene codes for a molecule called GRK3 that is involved in turning off the effects of another molecule (CRF) responsible for the brain's anxiety response. If the gene in some individuals is producing less GRK3 than needed, the consequence might be a greater anxiety response, a result that corresponds to bipolar symptoms.

Another practical benefit of genomics is pharmacogenics which is the development of pharmaceuticals based on knowledge of the human genome. Such drugs might supply a necessary protein to a patient who has a defective copy of the gene that should code for the manufacture of that protein. Or one of these drugs might be customized to help a patient with a particular genotype or set of gene variations. For example, the drug Herceptin, developed by Genentech, is designed to treat breast cancer in women who express the HER2 gene.

UNDERSTANDING HUMAN NATURE

The human genome is useful not just for medical applications but also as a guide to understanding human nature. Only 20,000 to 25,000 genes are responsible for coding all human proteins, yet they have a tremendous impact on human nature, from the physical to the psychological. Scientists have found, for example, two families of genes

Above: New pharmaceuticals are being developed based on increasing knowledge of the human genome. Top left: A computer image of a human DNA sequence. In 2003, the Human Genome Project announced it had mapped all three billion DNA letters in the human genome.

THE GENOME RACE

Begun in 1990, the Human Genome Project was an international program to sequence the entire human genome. It was a public effort whose U.S. component was the National Human Genome Research Institute (NHGRI), part of the federal government's National Institutes of Health. But beginning in 1999, private enterprise got into the game, courtesy of Celera Genomics. The company, headed by J. Craig Venter, began its own initiative to sequence the human genome. Celera used what its representatives called a "shotgun" sequencing technique, which was ultimately proven to be faster than the more traditional approach adopted by NHGRI. The idea that a private entrepreneur working for profit would challenge such a large-scale public initiative provoked criticism and concern.

In the end, both initiatives finished their first survey of the genome at the same time. In fact, a Harvard research team found that the rivalry spurred and improved the public project and left scientists with two valuable sequences to study.

The Human Genome Project, begun in 1990, was finished in 2003. Here, Francis Collins, director of the Human Genome Research Institute, announces its completion.

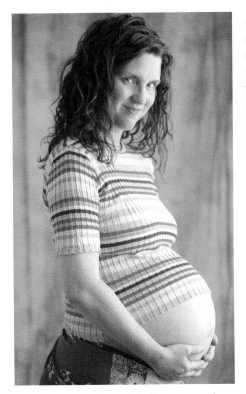

The extended gestation period of humans may be related to two families of genes that were duplicated in the human genome about 75 million years ago.

that were duplicated in the human genome after our divergence with rodents about 75 million years ago. These families encode sets of proteins that may be involved in the extended period of pregnancy special to humans.

Even personality traits may be governed by genes. Researchers in Israel have found an association between novelty seeking and the dopamine receptor-4 (DRD4) gene, a gene that codes for a molecule that receives dopamine, a neurotransmitter or messenger between neurons. In 2005, scientists reported a link between risk taking and neuroD2, a gene related to the development of the part of the brain called the amygdala.

OTHER GENOMES

The human genome is only one set of genes of interest to scientists. Every year, the genomes of other organisms are sequenced—plant, animal, and other. They help shed light not only on their own species but also on how the human species evolved and what makes us different from other animals. The mouse genome was sequenced in 2002. Researchers found that mice and humans share about 99 percent of their genes, which supports the usefulness of mice as models for disease in humans.

Other organisms whose genomes have been sequenced so far include the fruit fly, chimpanzee, chicken, rat, flu virus, dog, water flea, flowering plant *Arabidopsis thaliana*, and many bacteria.

Brave New World

The possibility of changing human nature through biological tinkering has long been both a dream and a nightmare. In science fiction works such as Aldous Huxley's *Brave New World*, writers have explored the bright and dark sides of expanding the human life span and altering the human genome. Now biological advances are making some of these possibilities more real— and renewing the ethical and practical questions they raise.

Genetic engineering—the technological manipulation of the genes of an organism—has already given rise to crop strains and animal breeds that may have never before existed in nature (although according to a theory known as horizontal transmission, virtually all genetic combinations, even between unrelated species, may potentially occur). The possibility that such techniques could be used to design "improvements" to human populations—greater height, intelligence, or health— brings with it the potentiality for monstrous mistakes and the threat of even greater human inequality than already exists. Other areas of study rife with both promise and peril include greater longevity and the creation of clones.

LONGER LIVES

Only several years ago, American biochemist Stephen Spindler discovered that elderly mice put on a low-calorie diet lived about six months longer than the typical mouse life span of two years. His oldest mouse lived for nearly five years—winning Spindler an award, the Methuselah Mouse Rejuvenation Prize, or "Mprize," created to encourage research into aging.

Above: A grower inspects corn. New crop strains have become possible through the use of genetic engineering, leading to increased production to feed the world's population. Top left: A rat in a Skinner Box, a device used to measure various types of animal response.

Spindler's research was just one of many attempts in the first decade of the twenty-first century to extend the span of life. American researchers found a way to extend the life span of yeast cells by up to 80 percent through the use of a group of compounds found in red wine and vegetables. The same molecules are also active in cultured human cells, giving hope that the method will have applications for extending human longevity.

Aubrey de Grey, scientist and cofounder of the Methuselah Foundation, granter of the Mprize, says he believes that the human life span could be extended by hundreds of years by the middle of the twenty-first century. But even a more modest extension of the human life span could have serious social consequences, as older generations fail to make way for younger ones, overcrowding increases, and age-related health problems proliferate.

CLONING

In August 2005, Korean scientist Hwang Woo-suk reported the first successful cloning of a dog. His team had taken DNA from the ear of an adult male Afghan hound, added it to a harvested egg, and successfully grown a clone, or genetically identical copy, of the DNA donor. The clone was named Snuppy, for Seoul National University Puppy. The feat was yet another advance in the art of cloning mammals, a significant one because canine eggs are difficult to work with.

Hwang's feat was colored by a scandal that arose over several charges, including the assertion that he lied about his accomplishments in human embryonic stem cell cloning. But the report about the dog cloning stood up to scrutiny and was one of several advances in cloning in the 2000s. However, the importance of this accomplishment has been severely tainted by the knowledge that virtually all of the other research from Hwang's lab is fraudulent. The dog cloning followed the cloning of a cat, announced in 2002, a feat that allowed a business called Genetic Savings and Clone to offer to clone beloved pet cats for $50,000 apiece. Rats, less valued as pets but highly valued as laboratory animals, were first cloned in 2003.

These achievements raised the possibility that humans might one day be cloned. Technical obstacles continue to block the way, along with laws forbidding the practice. But if human cloning ever does take place, it will raise ethical questions—including the ethics of deliberately giving someone a start in life as a genetic copy of a parent, rather than a mix of two parents' genes.

The world's first cloned cat is held by two doctors at the College of Veterinary Medicine at Texas A&M University on February 8, 2002. Scientists have yet to complete succesful cloning of a human embryo.

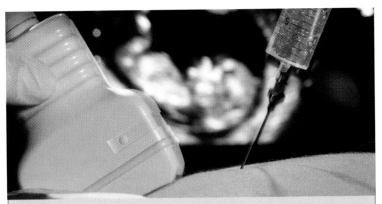

A needle draws amniotic fluid from the uterus of a pregnant woman, in the process of amniocentesis. The fluid will be analyzed for possible genetic or chromosomal disorders.

DESIGNER BABIES

Until recently, the genetic makeup of babies has always been determined the old-fashioned way—by the random shuffling of parents' genes in the act of sex. But increasingly, biology is making it possible to determine children's genetic makeup with greater precision. That possibility has raised questions about whether such "designer baby" practices are ethical—but the techniques are already taking shape.

One technique is amniocentesis, in which fluid from an expectant mother's womb is tested for genetic markers. The test is often used to determine whether the fetus has genetic abnormalities, giving the mother the option to abort the fetus. In China and India, it is often used to ensure that the child will be a boy.

Another technique, called preimplantation genetic diagnosis, allows parents to select a preferred embryo from among several conceived in a lab. Another technique—not yet approved for humans but often used with animals—is to use genetic engineering to alter the genes of an embryo so that it exhibits desired traits.

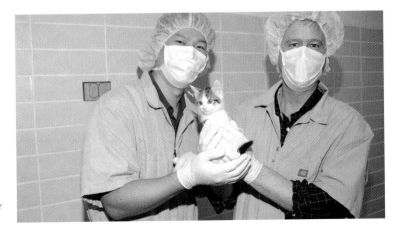

Penetrating the Wild

In 2005, Japanese scientists released the first photographs ever taken of a live giant squid in the wild. The photos of the 25-foot- (7.6-m) long cephalopod offered valuable information about how the enormous animals, until now veiled in mystery, swim and hunt in the ocean depths. It was one of many recent discoveries biologists have been making about plants, animals, and other organisms in the wild and closer to home.

NEW SPECIES

Biologists discover many new species every year. In November 2004, marine scientists reported that they had discovered 106 new species of fish in the past year alone, along with hundreds of new species of plants and other animals. The discoveries included, in the waters of Guam, a gold-speckled and red-striped goby fish that lives in partnership with a shrimp at its tail. Also at sea, Japanese scientists recently discovered two new species of rorqual whales, a group of long-bodied whales that also includes the common minke whale and the blue whale.

New species discoveries have also been taking place on land. In 2005, a scientist in Laos discovered a ratlike animal known by locals as kha-nyou. Believed to be a nocturnal, forest-dwelling vegetarian, the new species, *Laonastes aenigmamus*, represents an entirely new family of rodent. That same year, researchers in Africa found evidence of a strange population of apes that resemble giant chimpanzees but nest on the ground like gorillas. Whether they are a distinct culture of chimpanzees or a new subspecies remains unclear.

DISCOVERIES ABOUT EXISTING SPECIES

Biologists have also been making discoveries about the physiology and behavior of known species. Chimpanzees have long been

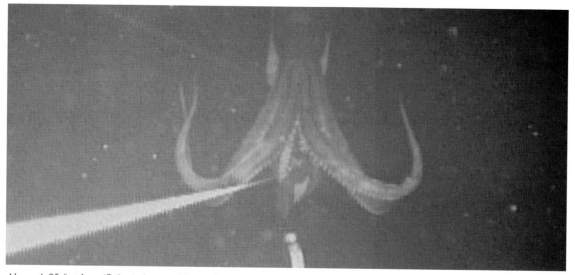

Above: A 25-foot-long (7.6-m) giant squid was photographed for the first time near Japan's Bonin Islands in 2005. Biologists have recently made a number of such new discoveries about previously unknown organisms that share our Earth. Top left: A ratlike animal discovered in Laos in 2005, named Laonastes aenigmamus, *represents an entirely new family of rodent.*

known to use tools for such purposes as getting food. But a 2005 study showed that they also develop cultural traditions of tool use by imitating one another. This phenomenon was observed when two groups of laboratory chimps developed different traditions of how to retrieve an item of food stuck behind a blockage in a system of tubes—one using a stick, the other poking without a stick. The study shed light on how human culture might have originated.

On a much smaller scale, researchers in 2004 discovered that the parasitic protozoan *Trypanosoma cruzi*—which causes the lethal Chagas' disease and affects the heart, colon, and esophagus—has an unusual way of propagating itself. Bits of its DNA become integrated into the DNA of an infected person. This disturbing discovery represents the first time that parasitic DNA has been encountered in the human genome.

ENVIRONMENTAL CHANGE

Biologists have also been finding out how organisms are reacting to environmental change, from global warming to pollution. In 2002, a study found that the American pika, a small mammal that lives in high mountains, is rapidly disappearing as its cold climate warms. It is one of many species suffering from the effects of human-induced climate change. On the other hand, other organisms are benefiting from human changes to the environment. It is known, for example, that bacteria can adapt to use noxious substances as nutrients, including chlorinated solvents present in dry-cleaning fluids, radioactive compounds, and petroleum.

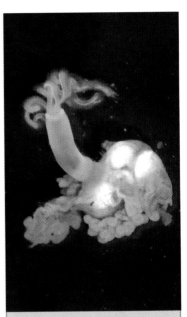

A female Osedax frankpressi *whale worm was discovered in 2002. It feeds on the bones of dead whales.*

WHERE THE SUN NEVER SHINES

In 2004, an entirely new genus of organism was discovered at the bottom of the ocean, where the Sun never shines. The organisms were two species of worms that feed on the bones of dead whales. The worms were unlike any other known animal in both body form and feeding strategies. They had no mouths, stomachs, eyes, or legs, but they did have red, feathery crowns, which turned out to be gills, and green structures that anchor them to the whale bones like roots. They use the roots to infiltrate the bones and digest the fats and oils inside them with the help of symbiotic bacteria. The worms were so unusual as to be classified in a new genus, *Osedax*, or "bone devourer."

Chimpanzees have long been known to use tools, but a 2005 study showed that they also develop cultural traditions of tool use by imitating one another. Such findings expand our understanding of the origins of human culture.

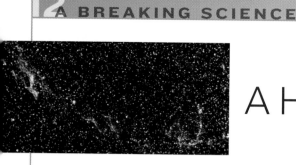

A Host of Discoveries

Above: U.S. soldiers wearing masks to protect against the influenza pandemic known as The Spanish Flu, which killed as many as 50 million people from 1918-1919. Top left: Exploring the possibility of life existing elsewhere in the universe is the work of exobiologists.

Many of the discoveries biologists make are hard to classify. They combine the expertise of different fields, or reach into areas so unfamiliar as to seem truly strange. Exobiologists, for example, specialize in exploring the possibility of life on other worlds, searching our solar system and distant stars for signs of life.

Exobiologists have yet to produce irrefutable evidence of life beyond Earth. But in other respects, biologists have had tremendous success in doing what seemed impossible. They have created a new kind of organism and resurrected a virus. And they discovered that in surprisingly recent evolutionary times, Homo sapiens shared its planet with an intelligent species nicknamed "hobbits."

CREATING LIFE
Ever since Mary Shelley published the horror novel *Frankenstein* in 1818, the idea of creating life has been a powerful one. In 2004, American chemist Peter Schultz and his team announced that they had created a form of life that is at least partially human-made—containing building blocks never before seen in nature.

The organism was derived from the bacterium *Escherichia coli;* the creation has a genetic code

based on 22 amino acids. Nearly all other living things on Earth, including humans, have a genetic code that uses only 20 amino acids to build proteins. But the new bacterium builds proteins out of 22: the naturally occurring 20, plus 2 unnatural ones.

RESURRECTING LIFE
Almost as remarkable as creating life is resurrecting it. In Michael Crichton's 1990 novel *Jurassic Park* and the movie that followed it, scientists resurrected dinosaurs by using recovered DNA to grow new and dangerous specimens. In 2005, American scientist Jeffrey Taubenberger and his colleagues did virtually the same thing with the influenza virus that killed 20 million or more people during the Spanish flu pandemic of 1918.

Although the virus had been lost to time, the scientists were able to resurrect it from DNA gathered from the preserved tissue of people who died in the pandemic. The hope is that this re-created virus will teach scientists how to avoid another pandemic; the fear is that it may get loose from the lab and cause a pandemic of its own.

LITTLE PEOPLE
Paleontologists are always discovering extinct species, but every once in a while they find one that captures the public imagination. That is what happened in 2004, when scientists reported on the remains of what appeared to be an ancient species of dwarf hominid. Known as *Homo floresiensis,* for the

Indonesian island of Flores on which they lived, these people stood just over three feet (.9 m) tall, earning them the nickname "hobbits." Yet they had enough intelligence to use fire and make sophisticated tools.

Until this discovery, scientists thought that *Homo sapiens* had been the sole species in the genus *Homo* since about 30,000 years ago, when the last Neanderthal died. But the remains of these little people dated to as recently as 18,000 years ago. There were even hints that they survived later. Local folklore described them as living there until the nineteenth century. The idea that not so long ago we shared the world with people who were human but not quite human was one of those startling surprises that only biology can provide.

In 2004, scientists reported the discovery of Homo floresiensis, *a cast of whose skull is shown at right. A "Hobbit-like" species is seen as a new link in the evolution of man.*

A feral hog, killed in 2004 by Chris Griffin, a Georgia resident was nicknamed "Hogzilla." Claims of its monstrous proportions turned out to be somewhat exaggerated.

CRYPTOZOOLOGY

Biology meets mythology in the field called cryptozoology. This is the search for organisms that most people presume to be nonexistent, such as the legendary Scottish lake dweller, the Loch Ness monster. Other famous examples include Ogopogo, a Loch Ness counterpart that may dwell in Lake Okanogan in British Columbia, Canada, the apelike Abominable Snowman (or Yeti) of the Himalayas, and his North American cousin, Bigfoot (or Sasquatch). Cryptozoologists use various methods, some of them scientific, to look for such beasts, often receiving little but mockery from the world.

Every once in a while, a cryptozoological claim makes the headlines. In 2004, reports surfaced from Georgia of a monster hog called Hogzilla, 12 feet (3.7 m) long and 1,000 pounds (454 kg). It had recently been killed by local resident Chris Griffin, who had a photograph to prove it. Had Hogzilla been real? A team of *National Geographic Explorer* experts put the story to rest when they dug up Hogzilla's body and found it had been only about 7.5 feet (2.3 m) long and 800 pounds (363 kg)—big, but not of the reported monstrous proportions.

A Sea of Controversy

Biology has been a controversial science ever since Vesalius contradicted the revered anatomical teachings of Galen in the sixteenth century. Since then, the social and political conflicts over the discoveries and techniques of biologists have only continued. The sea of controversy ranges from evolution to embryonic stem cells to environmentalism.

EVOLUTION IN THE SCHOOLS

Ever since Darwin first proposed the theory of evolution, this train of thought has met with religious opposition from creationists, who reject it because they think it contradicts the biblical account of creation. In the United States, creationists have long tried to stop evolution from being taught in public schools. In the Scopes trial, in 1925, Tennessee biology teacher John Scopes was convicted for breaking a state law forbidding the teaching of evolution. But by the second half of the century, the courts ruled repeatedly in favor of permitting the teaching of evolution in the schools, saying that creationism was a religious viewpoint and therefore had no place in determining public school curricula. Yet the creationist challenges to the teaching of evolution went on.

One such challenge came in 2004, when the Dover, Pennsylvania, school board adopted a policy requiring students to hear a statement skeptical of evolution. The statement said that evolution is "not a fact" and referred the students to a textbook on intelligent design, the view that complex organisms could have arisen only through the work of an intelligent designer. In 2005, a federal judge barred the policy, ruling that intelligent design "is a religious view, a mere re-labeling of creationism, and not a scientific theory."

STEM CELL RESEARCH

The harvesting of stem cells from embryos is another area of controversy. Many scientists believe that stem cells, cells with the ability to develop into any body tissue, can one day be used to replace damaged tissues and treat diseases. But mining the cells from embryos requires the destruction of the embryos, and many people think that this is murder. Often on religious grounds, they believe that a human embryo is a human being with a right to life, and therefore want to stop embryonic stem cell research. People who

Above: In a 1925 trial, Tennessee biology teacher John Scopes was found guilty of teaching the theory of evolution. Top left: In May, 2005, Jonathan Wells, a proponent of intelligent design, spoke with Connie Morris of the Kansas State Board of Education. The courts banned teaching of intelligent design in that state, calling it "a religious view ...not a scientific theory."

do not share their view want to permit it so potentially life-saving medical research can go on.

In 2001, President George W. Bush split the difference between the two camps, authorizing federal funding for embryonic stem cell research, but only for research on 60 existing stem-cell lines. The compromise did not satisfy scientists, who needed more lines to work with, nor did it satisfy people absolutely opposed to any embryonic stem cell research. The debate continues.

ENVIRONMENTAL CONFLICTS

Ecologists and other biologists involved in studying the environment are often active in trying to protect it. For example, they not only research biodiversity—the variety that exists among organisms—but try to call the public's attention to the ongoing decline in biodiversity.

Biological research has been important in showing that Earth is getting warmer. For example, in 2003, American researchers Camille Parmesan and Gary Yohe reported that they found effects of rising temperatures on 84 percent of the 334 species they studied, from longer growing seasons to changes in the location of migrations.

This research, too, has been controversial. There are those who are attempting to debunk

prevailing theory. These climate contrarians argue against the majority of scientists who believe there exists human-induced global warming. In 2001, Danish political scientist Bjørn Lomborg published *The Skeptical Environmentalist*, in which he argued against many environmentalist claims. Scientists rebutted his arguments. By 2004, however, *Time* magazine was calling him one of the world's 100 most influential people.

Above: The migration patterns and life cycles of many species of birds are threatened by climate change. Top: Stem cells from a mouse embryo. Human embryonic stem cells carry potentially lifesaving benefits. Some people, often on religious grounds, oppose stem-cell research.

GLOSSARY

A

ADAPTATION A change in a characteristic of an organism that better suits it to its environment.

AMINO ACID Any organic acid of which proteins are mostly comprised.

ANATOMY The study of the structure of organisms, especially of their internal parts.

ATP (ADENOSINE TRIPHOSPHATE) A molecule whose breakdown releases the energy necessary for the operations of cells.

B

BIODIVERSITY The biological variety existing among organisms and their environments.

BIOME A large geographic region characterized by a particular climate and dominant organisms.

BIOSPHERE The total region on or near Earth's surface where life exists.

BOTANY The study of plant life.

C

CELL The most basic biological unit, which contains organelles and which is capable of activities such as reproduction, growth, and metabolism. Cells can live on their own or as part of a multicellular (many-celled) organism.

CELLULOSE The main supporting material of plants and the main constituent of their cell walls.

CHLOROPHYLL A green pigment that can absorb solar energy. Plants use it to carry out photosynthesis, the conversion of sunlight to food.

CHROMOSOME A long, threadlike structure in the cell nucleus that contains DNA and carries the organism's hereditary information.

CIRCULATORY SYSTEM A group of organs that transports materials throughout an organism.

CLONE An organism that has exactly the same genes as another organism.

COMMUNITY A group of populations, including animals and plants that live and interact in the same area.

D

DIGESTIVE SYSTEM A group of organs that breaks down food into a form that can be used by the cells of an organism.

DNA (DEOXYRIBONUCLEIC ACID) The molecule also called nucleic acid that makes up an organism's genes and is found in chromosomes; DNA is a type of molecule called a nucleic acid.

E

ECOLOGY The study of the interactions between living things and their environment.

ECOSYSTEM A community of living organisms and their environment.

EMBRYO An organism in an early stage of development.

ENDOCRINE SYSTEM A group of organs that regulates body processes by secreting hormones into the blood.

ENVIRONMENT An ecological community.

ENZYME A molecule that speeds up chemical reactions in living things without being changed itself.

EUKARYOTE A multi-or single-celled organism in which the cell's genetic material is enclosed within a specialized membrane.

EVOLUTION The naturally occurring process of change in inherited characteristics by which one form of life gives rise to another.

EXCRETORY SYSTEM A group of organs that eliminates the waste products that arise as a result of an organism's metabolic activity.

EXOSKELETON A rigid, external framework, found in animals such as insects, that provides protection, support, and points of attachment for muscles.

F

FOOD CHAIN See food web.

FOOD WEB The network of organisms in an ecosystem that can eat or be eaten by others; a simplified version is a food chain, which represents the flow of energy from the sun to green plants to animal consumers.

G

GENE A unit of heredity consisting of a segment of a DNA molecule in a chromosome; the gene contains instructions for making a protein.

GENETIC DRIFT Random changes in the frequency of genes that do not tend to improve adaptation.

GENETIC ENGINEERING The manipulation of the genetic structure of organisms through technological means, such as the directed insertion of a gene(s) from one organism into the genetic material of another.

GENOME The entire set of genes carried in each of an organism's cells.

H

HEREDITY Transmission of traits from parent to offspring via genes.

HETEROTROPH An organism that cannot produce its own food, instead taking it in from other organisms; all animals are heterotrophs.

HOMEOSTASIS The ability of an organism to regulate itself so as to remain internally stable.

I

IMMUNE SYSTEM A group of organs that defends an organism against invaders.

INVERTEBRATE An animal that lacks a backbone.

M

MEIOSIS A form of nuclear division in which a cell divides into four daughter cells, each with only half the number of chromosomes in the parent cell.

METABOLISM The sum of the chemical reactions by which an organism's cells perform their activities, including both the breaking down (catabolism) and building up (anabolism) of molecules.

MICROBE An organism so small it can only be seen using a microscope; also known as microorganism.

MITOSIS A form of nuclear division in which the nucleus of a cell divides and forms two identical nuclei.

MUSCULOSKELETAL SYSTEM A group of muscles and bones attached in such a way as to permit locomotion (movement from place to place).

MUTATION A sudden alteration in genetic material; mutations naturally occur at a low rate and may happen faster in the presence of radiation and certain chemicals.

N

NATURAL SELECTION A process of evolution by which genes that increase the fitness of an organism are more likely to be inherited, leading to the modification of the species.

NERVOUS SYSTEM A body-wide group of receptors and transmitters that receive and analyze information from sensory organs and transmit instructions to effectors, such as muscles and glands.

NICHE The way of life or role of a species in its environment.

NUCLEIC ACID See DNA.

NUCLEUS The membrane-enclosed space in a eukaryotic cell that contains the genetic material.

O

ORGAN A part of an organism where two or more kinds of tissues perform a specific function.

ORGAN SYSTEM A group of organs that carries out a major activity in an organism.

ORGANELLE A structure within a cell made up of molecules and having a specific task.

ORGANIC COMPOUND A compound that contains carbon in combination with one or more other elements.

ORGANISM An individual living system capable of basic biological activities such as reproduction, growth, and metabolism.

P

PHOTOSYNTHESIS The process of conversion of sunlight into food; see also chlorophyll.

PHOTOTROPH An organism that can make food from sunlight through the process of photosynthesis; phototrophs include plants, algae, and certain bacteria.

PHYSIOLOGY The study of how organisms function.

POPULATION A group of organisms of the same species living in one area at one time.

PROKARYOTE An organism whose genetic material is not contained within a cell nucleus; bacteria are one example of a prokaryote.

PROTEIN A large molecule formed out of smaller molecules called amino acids; proteins are the principal building blocks of cells.

PROTIST An organism that has a cell nucleus and is usually microscopic and one-celled.

R

REPRODUCTIVE SYSTEM A group of organs that is involved in the creation, growth, and delivery of offspring.

RESPIRATORY SYSTEM A group of organs that provides living things with oxygen from the environment and disposes of waste products such as carbon dioxide.

S

SPECIES The basic unit of classification of living things; members of one species usually cannot interbreed with members of another.

SYMBIOSIS An interaction in which members of two different species live together in a close relationship for the benefit of at least one member.

T

TAXONOMY The study of the classification of organisms according to their natural relationships; taxonomy is also called systematics.

TISSUE A group of cells that share a similar structure and work together to perform a particular function.

V

VERTEBRATE An animal that has a backbone.

Z

ZOOLOGY The study of animals.

FURTHER READING

BOOKS

Asimov, Isaac. *Asimov's Chronology of Science & Discovery,* rev. ed. New York: Collins, 1994.

A Dictionary of Biology, 5th ed. Oxford: Oxford University Press, 2004.

Campbell, Neil A., and Jane B. Reece. *Biology, 7th ed.* San Francisco: Benjamin Cummings, 2004.

Dawkins, Richard. *The Blind Watchmaker: Why the Evidence of Evolution Reveals a Universe without Design.* New York: W.W. Norton & Co., 1986.

Edwards, Gabrielle I. *Biology the Easy Way, 3rd ed.* Hauppage, N.Y.: Barron's Educational Series, 2000.

Garber, Steven Daniel. *Biology: A Self-Teaching Guide, 2nd ed.* New York: Wiley, 2002.

Goldsmith, Timothy H., and William F. Zimmerman. *Biology, Evolution, and Human Nature.* New York: John Wiley & Sons, 2001.

Gould, Stephen Jay. *Wonderful Life: The Burgess Shale and the Nature of History.* New York: W.W. Norton & Co., 1989.

Hine, Robert, ed. *The Facts on File Dictionary of Biology, 3rd ed.* New York: Facts on File, 1999.

Janovy, Jr., John. *On Becoming a Biologist.* New York: Harper & Row, 1985.

Keller, Rebecca. *Real Science-4-Kids Biology, Level I.* Albuquerque, N.M.: Gravitas Publications, 2005.

Layman, Dale. *Biology Demystified: A Self-Teaching Guide.* New York: McGraw-Hill, 2003.

Mader, Sylvia S. *Inquiry into Life, 11th ed.* New York: McGraw-Hill, 2004.

Margulis, Lynn, and Karlene V. Schwarz. *Five Kingdoms: An Illustrated Guide to the Phyla of Life on Earth, 3rd ed.* New York: W.H. Freeman and Co., 1999.

Mayr, Ernst. *This Is Biology: The Science of the Living World.* Cambridge, Mass.: Belknap Press, 1997.

McGrath, Kimberley A., ed. *World of Biology.* Farmington Hills, Mich.: Gale Group, 1999.

Raven, Peter H., et al. *Biology, 7th ed.* New York: McGraw-Hill, 2004.

Ridley, Matt. *Genome: The Autobiography of a Species in 23 Chapters.* New York: Harper Perennial, 2000.

Serafini, Anthony. *The Epic History of Biology.* New York: Plenum Press, 1993.

Siegfried, Donna Rae. *Biology for Dummies.* New York: Hungry Minds, 2001.

Southwood, Richard. *The Story of Life.* Oxford: Oxford University Press, 2003.

VanCleave, Janice. *Janice VanCleave's Biology for Every Kid: 101 Easy Experiments That Really Work.* San Francisco: Jossey-Bass, 1990.

WEB SITES

Action Bioscience | www.actionbioscience.org
Education resource connecting biology to pressing issues such as biodiversity and biotechnology.

Biology Century |
mywebpages.comcast.net/biologycentury
A summary of biological knowledge built around major concepts.

Biology-Online | www.biology-online.org
Tutorials, dictionary, forum, and online directory on biology.

The Biology Project | www.biology.arizona.edu
Online interactive resource for learning biology.

Cracking the Code of Life |
www.pbs.org/wgbh/nova/genome/program.html

Two-hour video with background information on the human genome.

Ecology.com | www.ecology.com
Information, news, and links on ecology.

Human Anatomy Online |
www.innerbody.com/htm/body.html
Interactive tour of aspects of human anatomy.

Human Genome Project Information |
www.ornl.gov/sci/techresources/Human_Genome/
home.shtml
Facts about the Human Genome Project and related issues.

Kimball's Biology Pages |
users.rcn.com/jkimball.ma.ultranet/BiologyPages/
Online biology textbook by biology professor John W. Kimball.

Living Things | www.fi.edu/tfi/units/life
Information from the Franklin Institute Online on the world's plants, animals, and other living things.

Microbes.info: The Microbiology Information Portal |
www.microbes.info
Portal to information on microbes, microbiology, and microbial diseases.

Online Biology Book Palaeos: The Trace of Life on Earth | www.palaeos.com
Information about life in past geologic ages.

Understanding Evolution | evolution.berkeley.edu
Basic facts and latest news about evolution.

MAGAZINES

Discover | www.discover.com

National Geographic | www.nationalgeographic.com

Natural History | www.naturalhistorymag.com

New Scientist | www.newscientist.com

ScienceDaily | www.sciencedaily.com

Science Magazine | www.sciencemag.org

Scientific American | www.sciam.com

Smithsonian | www.smithsonianmagazine.com

ORGANIZATIONS

American Association for the Advancement of Sciences | www.aaas.org

American Institute of Biological Sciences | www.aibs.org

American Museum of Natural History | www.amnh.org

Federation of American Societies for Experimental Biology | www.faseb.org

Field Museum of Natural History |
www.fieldmuseum.org

Linnaean Society of London | www.linnean.org

Marine Biological Laboratory (Woods Hole, Massachusetts) | www.mbl.edu

National Academy of Sciences | www.nasonline.org

National Museum of Natural History, Smithsonian Institution | www.mnh.si.edu

Natural History Museum (London) | www.nhm.ac.uk

Sierra Club | www.sierraclub.org

Society for Conservation Biology | www.conbio.org

Society for In Vitro Biology | www.sivb.org

WWF International (formerly known as the World Wildlife Fund) | www.panda.org

AT THE SMITHSONIAN

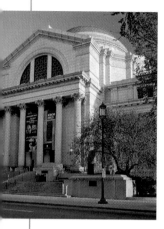

Biology is part of what the Smithsonian Institution is all about. Two of the institution's facilities, the National Museum of Natural History and the National Zoo, provide a wealth of resources for anyone interested in learning about living things—how they are built, how they function, and how they interact.

With its fossils and preserved organisms, the National Museum of Natural History allows you an upclose look at the anatomical details of organisms. At the National Zoo, you can observe living things in action.

Above: The Triceratops at the Natural Museum of Natural History is the world's first-ever mount of this plant-eating dinosaur from the Cretaceous Period, more than 65 million years ago. The bones of the Triceratops, which first went on display in 1905, had been a composite of the bones of at least ten different animals. Recently, however, advances in technology have allowed conservators to create an anatomically correct skeleton made of resin, plaster and fiberglass. There still is no known complete skeleton of Triceratops anywhere. Far left: The front entrance to the National Museum of Natural History.

The National Museum of Natural History
10th Street and Constitution Ave., NW
Washington, D.C. 20560
202-633-1000
info@si.edu
http://www.mnh.si.edu

The National Museum of Natural History has permanent collections and changing exhibits on many aspects of life, literally from A (anthropology) to Z (zoology). Are you interested in the study of living plants (botany), fossils (paleobiology), and insects (entomology)? The museum is the place to be. Among the museum's exhibits that are particularly relevant to this book are:

DINOSAUR HALL

To make evolutionary history come alive (chapter eight), tour the museum's extensive collection of fossils, especially those of the dinosaurs. They may be extinct, but they are always a popular stop,

and the museum's Dinosaur Hall gives them their due. A recent addition is the Triceratops exhibit, which includes a Triceratops skeleton replica and a virtual-motion model that uses digital technology to show how this behemoth moved.

THE HALL OF MAMMALS

The Kenneth E. Behring Family Hall of Mammals allows you to delve deeper into many areas discussed in this book—mammal evolution (chapter eight), present-day mammals (chapter ten), and mammals as part of ecosystems (chapter seven). The hall includes hundreds of preserved animals, displayed in their natural environments, from polar to desert regions, with many interactive displays. Wondering about a predator-prey relationship? A look at a leopard lounging in a tree high above you, with a dead impala stashed nearby for a later meal, will make the "eat-or-be eaten" lifestyle crystal clear.

A manatee suspended in air is one of the 274 taxidermied mounts on display in the Smithsonian's Hall of Mammals. What makes a mammal a mammal? To gain entry into this elite group that consists of more than 5,000 species, animals must share the following characteristics: hair, mother's milk, and three inner-ear bones.

The National Zoo
3001 Connecticut Ave., NW
Washington, D.C. 20008
202-633-4800
nationalzoo@si.edu
http://nationalzoo.si.edu

Spread out over an area that encompasses 163 acres, the National Zoological Park is home to more than 2,400 animals belonging to more than 400 species. Giant pandas are probably the museum's best-known residents, but many of the other residents are living and breathing examples of topics in this book.

CHEETAHS AND OTHER VERTEBRATES

The adaptations that have made cheetahs such speed demons (chapter six) are on display at the National Zoo. Several cheetahs, some born at the zoo, demonstrate their speed by chasing the lure, a flag attached to a cord, as it races around their enclosure.

Cheetahs are just one part of the ecology of the African savannah (chapter seven), and many other species from that biome are represented. African vertebrates at the zoo include lions, hippopotami, giraffes, gorillas, zebras, and gazelles. Other vertebrates at the zoo hail from such far-flung regions as Asia, Australia, North America, and Amazonia (chapter ten).

INVERTEBRATES AND PLANTS

In the wild, invertebrates are much more abundant than vertebrates (chapter ten), and the Invertebrate Exhibit has a great variety of them. They include giant African millipedes, spiny lobsters, anemones, leaf-cutter ants, cuttlefishes, and octopuses. Like all living things, their lives are intertwined with those of other organisms. This is especially clear in the Pollinarium, adjacent to the Invertebrate Exhibit.

The Pollinarium is a greenhouse dedicated to pollination, the process by which insects collect food while fertilizing plants (chapters seven and ten). Visitors to the Pollinarium watch as butterflies and bees visit flowering plants such as heliconia and red salvia. An entire colony of bees lives within a clear-walled enclosure that makes it easy and safe to watch the daily life of the hive.

INDEX

ACKNOWLEDGMENTS & PICTURE CREDITS

ACKNOWLEDGMENTS

The author thanks Alison Fromme, who contributed Chapters 7 and 9, and Melinda Corey, who contributed chapters 10 and 11. He also thanks Martha Corey-Ochoa.

The publisher wishes to thank Don E. Wilson, Smithsonian National Museum of Natural History; J'Nie Woosley and Ann Batdorf of the Smithsonian National Zoological Park; Ellen Nanney, Senior Brand Manager with Smithsonian Business Ventures; Katie Mann with Smithsonian Business Ventures; consultant Brian Hoyle; Collins Reference executive editor Donna Sanzone, editor Lisa Hacken, and editorial assistant Stephanie Meyers; Hydra Publishing president Sean Moore, publishing director Karen Prince, art director Edwin Kuo, designers Rachel Maloney, Mariel Morris, Gus Yoo, Greg Lum, La Tricia Watford, Erika Lubowicki, editorial assistant Aaron Murray, project editor Lisa Purcell, editors Marcel Brousseau, Molly Morrison, Suzanne Lander, Gail Greiner, Ward Calhoun, Emily Beekman, Liz Mechem, and Roger Ochoa, copy editors Glenn Novak and Eileen Chetti, picture researcher Ben Dewalt, production manager Sarah Reilly, production director Wayne Ellis, and indexer Jessie Shiers; Chrissy McIntyre of Chrissy McIntyre Research, LLC; Wendy Glassmire of the National Geographic Society; Harriet Mendlowitz of Photo Researchers, Inc.; and Crystal Smith of the National Library of Medicine.

PICTURE CREDITS

The following abbreviations are used: PR–Photo Researchers, Inc.; SPL–Science Photo Library; JI–© 2006 Jupiterimages Corporation; SS–ShutterStock; IO–Index Open; iSP–©iStockphoto.com; BS–Big Stock Photos; ARS/USDA–Agriculture Research Service/U.S. Department of Agriculture; NOAA–National Oceanic and Atmospheric Association; OAR–Oceanic and Atmospheric Research; NURP–National Undersea Research Program; USFWS–U.S. Fish and Wildlife Service; CDC–Centers for Disease Control and Prevention; NLM–Courtesy of the National Library of Medicine, SI/NZP–Smithsonian Institution/National Zoological Park; SIL–Smithsonian Institute Library; DL–Dibner Library Portrait Collection; SPS–Smithsonian Photographic Services; AP–Associated Press; LoC–Library of Congress; NGIC–National Geographic Image Collection
(t=top; b=bottom; l=left; r=right; c=center)

Introduction: Welcome to Biology
IV JI VI Eye of Science/PR 1t JI 1b JI 2bl JI 3l JI 3r SPL/Pasca Goetgheluck

Chapter 1: The Mystery of Life
4 IO/LLC, FogStock 5t IO/Keith Levit Photography 5b IO/Hot Ideas 6t Petit Format/PR 6bl G. Murti/PR 6br JI 7d IO 7tr NOAA/Frank and Joyce Borek 8 Biophoto Associates/PR 9 Illustration by Rachel Maloney 10d David T. Roberts/PR 10r iSP/Andrew Robinson 11bl ARS/USDA/Peggy Greb 11tr Eye of Science/PR 12d NOAA/OAR/NURP/Texas A&M University 12bl USFWS 13t NOAA/Carol Baldwin 13b JI 14d JI 14b Jessie Cohen/SI/NZP 15d Dr. Jeremy Burgess/PR 15tr iSP/Julie de Leseleuc 15br JI

Chapter 2: From Microscopes to CAT Scans
16 Jessica Bethke/SS 17t JI 17b JI 18tl ARS/USDA/Scott Bauer 18r ARS/USDA/Scott Bauer 19t ARS/USDA/Scott Bauer 19b ARS/USDA/Reith Weller 20d ARS/USDA/Jack Dykinga 20r CDC/James Gathany 21d Sheila Terry/PR 21tr ARS/USDA/Keith Weller 21b SPL/TEK Images 22d ARS/USDA/Peggy Greb 22tr iSP/Jibby Chapman 22bl iSP/Hallgrimur Arnarson 22br CDC/James Gathany 23 ARS/USDA/Eric Erbe 24d NOAA/OAR/NURP 24bl NOAA/OAR/NURP/R. Wicklund 24r ARS/USDA/Keith Weller 25l ARS/USDA/Jack Dykinga 25r AP/San Diego Union Tribune/John Gibbons 26d ARS/USDA/Sandra Silvers 26bl JI 26br ARS/USDA/Sandra Silvers 27 ARS/USDA/Keith Weller 28tl Mauro Fermariello/PR 28b Sheila Terry/PR 29bl ARS/USDA/Keith Weller 29tr Mauro Fermariello/PR

Chapter 3: Centuries of Scrutiny
30 Explorer/PR 31t NLM 31b NOAA 32d Courtesy of the New York Academy of Medicine Library 32bl JM Labat/PR 33bl NLM 33tl NLM 33tr NLM 34tl NLM 34br NLM 35tl Mary Evans/PR 35bl NLM 36tl NLM 36bl SIL 37tr LoC 37br J.L. Charment/PR 38d NOAA 38jaime Abecasis/PR 38t SIL 39 NOAA 40d iSP/Konstantinos Kokkinis 40b NLM 41bl SPL/PR 41tr LoC 42d iSP/cre8tive studios 42bl Darwin Dale/PR 42r NLM 43 A. Barrington Brown/PR 44d iSP/Tim Pleasant 44bl NLM 45 iSP/Stephen Sweet 45b Gusto/PR

Chapter 4: The Building Blocks of Life
46 Asa Thoresen/SPL 47t iSP/Monika Wisniewska 47b Steve Gschmeissner/SPL 48d Omikron/PR 48b iSP/Alex Dykes 49bl Charles D. Winters/SPL 49tr Travis Klein/SP 49br iSP/Milan Radulovic 50d Biology Media/PR 50b Biophoto Associates/PR 51d JI 51r Christopher Poliquin/SS 52d Biophoto Associates/PR 52b Russell Kightley/PR 53bl BS/Joss 53tr iSP/Craig Neltri 54d Torunn Berge/PR 54bl Steve Gschmeissner/SPL 54br eye of Science/PR 55 Steve Gschmeissner/PR 56d CNRI/PR 56bl Jennifer Waters and Adrian Salic/PR 56br Jennifer Waters/PR 57tr T.W./SS 57bl Jennifer Waters and Adrian Salic/PR 57br Jennifer Waters and Adrian Salic/PR

Chapter 5: A Complex Architecture
58 Mehaukulyls/PR 59t JI 59b Caroline K. Smith, MD/SS 60d Steve Gschmeissner/PR 60bl Caroline K. Smith, MD/SS 60br Mauro Fermariello/PR 61tr A. Pasieka/PR 61bl Mike Tolstoy/photobank.kiev.ua/SS 62d JI 62bl JI 62br JI 63d JI 63tr Dave Roberts/SPL 63bl JI 64d iSP/Matthew Cole 64b M.I. Walker/PR 65t © SJ Elmhurst BA Hons 2005/www.livingart.org.uk 65b JI 66d BS/Carolina Smith 66tr BS/Jyothi N. Joshi 66b iSP 67l BS/Lleha 67r Neil Borden/PR 68d Eric Grave/SPL 68cl JI 68cr Anatomical Travelogue/SPL 69l Parviz M. Pour/SPL 69r © 2006 Getty Images 70d JI 70r IO/photos.com select 71t Steve Gschmeissner/SPL 71b Rey Rojo/SS 72d iSP/Vera Bogaerts 72b Eye of Science/SPL 72br Professor Miodrag Stojkovic/SPL 73tr Francoise Sauze/SPL 73br iSP/Simon Webber

Chapter 6: Breathing and Eating
74 iSP/Robert Deal 75t iSP/Daniel Halvorson 75b Eye of Science/SPL 76d Pete Madison/SS 76b Dr. Kari Lounatmaa/SPL 77t iSP/Yorgos Arvanitis 78d IO 78b Susumu Nishinaga/SPL 79t Tan Kian Khoon/SS 79b Susumu Nishinaga/SPL 80d JI 80b David M. Martin, MD/SPL 81t Gordana Sermek/SS 81b Steve Gschmeissner/SPL 82d Steve Gschmeissner/SPL 82tr Morris Huberland/PR 82b Joan Ramon Mendo Escoda/SS 83bl Big Zen Dragon/SS 83r Alfred Pasieka/SPL 84d iSP 84bl Eric Florentin/SS 85t Des and Jen Bartlett/NGIC 85b iSP/Simon Mitchell 86d Chris Harvey/SS 86b CAMR/A.B. Dowsett/SPL 87tr Jaimie Duplass/SS 87c © Dattatreya/ Alamy 87bl © Thinkstock / Alamy 88d iSP/Alexandre Azevedo 88bl © MELBA PHOTO AGENCY / Alamy 88br © 2006 Getty Images 89tr JI 89br Dmitry/SS 90d Sherrianne Talon/SS 90bl Matt Antonino/SS 91d iSP/Steffen Foerster 91bl Paige Falk/SS 91br Arthur Ng Heng Kui/SS

Ready Reference
92t (l-r) SPL; SIL/DL; SIL/DL; Shaw House/www.shawprize.org; A. Barrington Brown/PR 92c (l-r) George Bernard/PR; LoC; SPL; SIL/DL 92b (l-r) SPL; Adam Hart-Davis/SPL; SPL; SPL; SIL/DL 93t (l-r) John Reader/PR; BIS/SL; SIL/DL; NLM/SPL; SPL 93c (l-r) SIL/DL; NLM; SIL/DL; SPL; SIL/DL 93b (l-r) George Bernard/PR; SIL/DL; A. Barrington Brown/PR; Gusto/PR; SPL 94l John Said/SS 94cl SIL/DL 94cr NLM 94t Wikimedia 94br Wikimedia 95bl SPL 95t SIL/DL 95br NLM 96d NLM 96bl SIL 96c NLM 97bl John M. Daugherty/SS 97bc Des Bartlett/SPL 97bc SPL 97tr iSP/Magnus Ehinger 97br National Human Genome Research Institute/The Broad Institute of MIT and Harvard 98l Jason T. Ware/PR 98t John Kirinic/SS 98br iSP/Nicola Stratford 99tr Richard Ellis/PR 99cr Richard Ellis/PR 99br Richard Ellis/PR 100 Illustration by Rachel Maloney/Data source: ©2006 Discovery Communications Inc. 101l (t-b) PR; phdpsx/SS; Yan Ke/SS; IPichugin Dmitry/SS, cGordon Snell/SS, riSP/Georg Hafner; CJ Photography/SS; Michael Almond/SS; lMichael Thompson/SS, rUlrike Hammerich/SS; photobar/SS; Steven Bourelle/SS 102d James Cavallini/PR 102tc M.I. Walker/PR 102tr William Attard McCarthy/SS 102c M.I. Walker/PR 102d iSP/SciMAT/PR 102bc Eye of Science/PR 102br AJPhoto/PR 103d Menna/SS 103tc Kerry L. Werry/SS 103tr Peter Guess/SS 103cl Anson Hung/SS 103c Joe Gough/SS 103cr Keir Davis/SS 103bl Elena Elisseeva/SS 103bc Christa DeRidder/SS 103br Julie Fine/SS 104 Carlyn Iverson/PR 105l Roger Harris/PR 105tr Jennifer Waters/PR 105br Brian Evans/PR

Chapter 7: The Community of Life
106 JI 107t JI 107b JI 108d zastavkin/SS 108r IO 109t Jan Erasmus/SS 109b JI 110d IO 110bl zasatvkin/SS 111bl JI 111tr IO/Keith Levit Photography 111br JI 112d M.I. Walker/PR 112bl Dwight Lyman/SS 112br Yuriy Maksymenko/SS 113d Melissa Dockstader/SS 113tr JI 114d JI 114br Gary Hincks/SPL 115t Paul Marcus/SS 115bl Chin Kit Sen/SS 116d Vilmos Varga/SS 116bl iSP/Kenneth C. Zirkel 117tr D.H. Snover/SS 117c Steve McWilliam/SS 117b Andrija Kova/SS 118d Gregory Dimijian/PR 118bl John M. Coffman/PR 119t Russel Swain/SS 119b David Scharf/SPL 120d JI 120r JI 120bl JI 121bl JI 121tr George Bernard/SPL

Chapter 8: The Evolving Tapestry
122 Bob Ainsworth/SS 123t JI 123b JI 124d Linda Bucklin/SS 124bl Vaide/SS 124br JI 125 Scott Bowlin/SS 126d JI 126bl Piotri Bieniecki/SS 127bl JI 127tr JI 128d Ng Yin Chem/SS 128tr SPL/PR 128b JI 129tr Richard Ellis/SPL 129b JI 130d JI 130b Jan Martin Will/SS 131d Mornee Sherry/SS 131bl iSP/Tamara Bauer 131tr Ritu Manoj Jethani/SS 132d JI 132b NOAA/P. Rona 133t Chase Studio/PR 133b Lynsey Allan/SS 134d JI 134tr JI 134bl James L. Amos/PR 135 Tom McHugh/PR 136d Chris Butler/SPL 136tr DK Pugh/SS 136bl Jacan/PR 137tr Joe Tucciarone/PR 137b Lawrence Lawry/PR 138d Chase Studio/PR 138r David R. Frasier/PR 139tr John Reader/PR 139bl Publiphoto/PR 139br Pascal Goetgheluck/PR

Chapter 9: What's in a Name?
140 JI 141t JI 141b Anita/SS 142 JI 142bl George Bernard/PR 142r Xtreme Safari, Inc./SS 143tr iSP/Romko_chuk 143br Tyler Olson/SS 144d Denis Pepin/SS 144bl LoC 144br Uwe Ohse/SS 145 JI 146d JI 146bl JI 147tl Newton Page/SS 147tr JI 147bl JI 147br Joao Estevao A. Freitas/SS 148d JI 148c GeoM/SS 149d Eye of Science/PR 149tr JI 149c Biophoto Associates/PR 150d Eric Grave/SPL 150tr Andrew Svred/PR 150b Ulrike Hammerich/SS 151t Gregory Ochocki/PR 151b Laurin Rinder/SS 152d JI 152br Clara Natoli/SS 153bl Astrid and Hanns-Frieder Michler/PR 153tc Mikhail Lavrenov/SS 153bc JI 153tr JI

Chapter 10: Plants and Animals
154 JI 155t JI 155b JI 156d IO/Mark Windom 156b David Roberts/Nature's Images/SS 157t IO/LLC, Fogstock 157b JI 158d Chris Harvey/SS 158b JI 159d Tracy/SS 159r JI 160d Brooke Barber/SS 160r JI 161tr Semen Lixodeev/SS 161cl JI 161bl Asther Lau Choon Siew/SS 162d JI 162bl Susumu Nishinaga/PR 162br Richard Marpole/SPL 163d JI 163tr JI 164d JI 164b JI 165d JI 165tr Linda Bucklin/SS 166d JI 166tr JI 166b JI 167tr JI 167cl JI 167cr JI 168d Patrick J. Lynch/PR 168tr Dante Fenolio/PR 168b JI 169l JI 169r JI 170d JI 170br IO/Mark Windom 171tr Andrew F. Kazmierski/SS 171bl LoC/Donald Hossack Bain

Chapter 11: The Human Animal
172 Hardas/PR 173tr IO/FogStock, LLC 173b JI 174d JI 174b JI 175t JI 175b JI 176d JI 176bl JI 176br JI 177l JI 177r Volker Steger/PR 178d JI 178c Sheila Terry/SPL 178tr Mehau Kulyk/SPL 179t JI 179b JI 180d JI 180bl JI 180r JI 181t BSIP, Gems Europe/SPL 181b JI 182d IO/Fogstock, LLC 182b JI 183t JI 183b IO/Fogstock, LLC 184d JI 184tr Graca Victoria/SS 184bl NLM/SPL 185bl JI 185r Jackie Lewin, Royal Free Hospital/SPL 186d JI 186bl JI 187d Conor Caffrey/SPL 187r Mikhail Levit/SS

Chapter 12: A Breaking Science
188 Hardas/PR 189t JI 189b JI 190d Francois Mori/AP 190tr Illustration by Mariel Morris/Data Source: AP/Dr. Maria Siemionow, the Cleveland Clinic 190br Amy Sancetta/AP 191 Robin Laurance/PR 192d Neil Borden/SPL 192b Simon Fraser, Hexham General/SPL 193t JI 193b JI 194d James King-Holmes/SPL 194b JI 195tr Evan Vucci/AP 195b JI 196d Walter Dawn/PR 196b JI IO/Fogstock, LLC 197tr Saturn Stills/SPL 197b Texas A&M University/AP 198d Wildlife Conservation Society/AP 198b National Science Museum/AP 199bl Kelly Shipp/SS 199tr © 2003 Greg Rouse 200d JI 200r SPL 201bl Richard Lewis/AP 201tr River Oak Plantation/AP 202d Topeka Capital Journal/AP 202b AP 203tr Andrew Paul Leonard/PR 203bl Jason Cheever/SS 203br David Kay/SS

At the Smithsonian
210l James DiLoreto/SPS [2003-8959] 210r D.E. Hurlburt, James DiLoreto/SPS [2001-4563.07] 211 James DiLoreto/SPS [2003-39147]